Exploring the Mid-Republican Origins of Roman Military Administration

This volume demonstrates the development of Roman military bureaucracy during the Middle Republic, expanding on recent research to examine these administrative systems that made possible Rome's expansion in this period.

Bringing together literary works, epigraphy, archaeology, topography and demography, the study reveals a complex and well-structured bureaucratic system developing in parallel with the army during the Middle Republic, propelled in no small part by the stresses of the Hannibalic War. Not only the contents of documents, but also the physical objects, individuals and spaces are discussed to re-create the administrative processes in maximum detail.

Exploring the Mid-Republican Origins of Roman Military Administration provides an invaluable resource for students and scholars of Rome's military and administrative history, as well as for anyone working on the Republican period.

Elizabeth H. Pearson is an independent scholar. She completed her PhD at the University of Manchester, United Kingdom, in 2016. In 2020, she won the Society of Military History's Vandervort Prize for her article 'Decimation and Unit Cohesion: Why Were Legionaries Willing to Perform Decimation?'.

Routledge Monographs in Classical Studies

For more information on this series, visit: https://www.routledge.com/
Routledge-Monographs-in-Classical-Studies/book-series/RMCS

Exploring the Mid-Republican Origins of Roman Military Administration

With Stylus and Spear

Elizabeth H. Pearson

Routledge
Taylor & Francis Group

LONDON AND NEW YORK

First published 2021
by Routledge
2 Park Square, Milton Park, Abingdon, Oxon OX14 4RN

and by Routledge
52 Vanderbilt Avenue, New York, NY 10017

Routledge is an imprint of the Taylor & Francis Group, an informa business

© 2021 Elizabeth H. Pearson

British Library Cataloguing-in-Publication Data
A catalogue record for this book is available from the British Library

Library of Congress Cataloging-in-Publication Data
Names: Pearson, Elizabeth H. (Historian), author.
Title: Exploring the mid-Republican origins of Roman military Administration : with stylus and spear / Elizabeth H. Pearson.
Description: Abingdon, Oxon ; New York, NY : Routledge, [2021] |
Series: Routledge monographs in classical studies | Includes bibliographical references and index. |
Identifiers: LCCN 2020045632 (print) | LCCN 2020045633 (ebook) |
ISBN 9780367820732 (hardback) | ISBN 9780367745547 (paperback) | ISBN 9781003014768 (ebook)
Subjects: LCSH: Military administration--Rome. |
Rome--Army--Organization. | Rome--History--Republic,
510-30 B.C. | Bureaucracy--Rome--History.
Classification: LCC U35 .P43 2021 (print) | LCC U35 (ebook) |
DDC 355.60937/09014--dc23
LC record available at https://lccn.loc.gov/2020045632
LC ebook record available at https://lccn.loc.gov/2020045633

ISBN: 978-0-367-82073-2 (hbk)
ISBN: 978-0-367-74554-7 (pbk)
ISBN: 978-1-003-01476-8 (ebk)

Typeset in Sabon
by MPS Limited, Dehradun

Contents

Figures

Tables

Acknowledgements

This book has been a long time in the making, and there are many people who deserve thanks for their help and support. Everyone who shared my time in the Classics and Ancient History Department at the University of Manchester is due thanks for making my time there so enjoyable and interesting, but there are far too many to name here. Likewise the delegates of the always-fascinating and much-missed International Ancient Warfare Conference. In particular, I would like to thank Andy Fear, Tim Cornell and Polly Low for their helpful comments and exhaustive discussions on this book in its thesis form, without which it would be much the poorer. I would also like to thank my examiners, Tim Parkin and Kate Gilliver, for their supportive comments and suggestions. Thanks also to April Pudsey for her comments on the demographic elements and to Nathan Rosenstein, Toni Ñaco del Hoyo and James Tan for their discussion of and constructive comments on *tributum*.

I would like to thank Routledge for publishing the book and all those involved in taking it from manuscript to finished article.

Finally, my heartfelt thanks go to my husband, my family and my friends, who have uncomplainingly served as sounding boards and proofreaders, provided countless cups of tea and generally supported me throughout what has sometimes been a very trying process. I couldn't have done it without you.

This thesis on which this book is based was funded by the Arts and Humanities Research Council and the University of Manchester's President's Doctoral Scholar Award.

Abbreviations

All abbreviations of ancient texts are given in the form provided by the *Oxford Classical Dictionary*[4].

CIL	*Corpus Inscriptionum Latinarum* (1863–), Berlin.
FRHist	Cornell, T. J. *et al.*, 2013. *The Fragments of the Roman Historians: Vols. 1–3,* Oxford.
LSJ	Liddel, H. D., Scott, R., Jones, H. S., 1925. *A Greek-English Lexicon: A New Edition,* Oxford.
OLD	Glare, P. G. W. (ed.), 1982. *Oxford Latin Dictionary,* Oxford.
RMR	Fink, R.O., 1971. *Roman Military Records on Papyrus,* London.
Tab. Vind.	Bowman, A. K., Thomas, J. D., 1994. *The Vindolanda Writing-Tablets (Tabulae Vindolandenses I–III),* London.

Introduction

Roman military bureaucracy began, at some point in the sixth century BC, with the census.[1] This document was created as a record of Roman military manpower and remained a regular feature of Roman administration throughout the Republic. In the fourth century AD, Vegetius' study of the army, *De re militari*, emphasised the importance of extensive bureaucracy to its successful functioning. Between the two, however, literary sources are lacking, creating something of an evidentiary desert. The situation has improved in recent decades, aided greatly by the continuing excavations at Vindolanda, a fort located on the Stanegate around a mile south of – and predating – Hadrian's Wall. Among the finds are military documents dating to the late first and early second centuries AD, which concern provisioning, accounting and the composition and disposition of units and their members.[2] Papyri discovered earlier at Dura-Europos, a fort on the Euphrates in Syria, also cover issues of provision, pay and deployment, primarily of the *cohors XX Palmyrenorum*.[3] These documents have been discussed in detail by a number of scholars, allowing much more to be learnt both about the functioning of the army as a whole and about the lives of soldiers in the Principate.[4] Overall, they support the picture put forward by Vegetius of an army made stronger and more effective by its complex bureaucracy.[5]

However, guided as it is by extant documents, the scholarly discussion of Roman military administration has been largely limited to the first century BC and the Principate.[6] Little has been written concerning the origins and development of military bureaucracy. This tendency tallies with the wider study of the Roman army. In Keppie's *The Making of the Roman Army*, a single chapter deals with the army from the founding of Rome in the eighth

---

1 Livy 1.42–4. Unless otherwise stated, all dates are BC.
2 E.g., especially *Tab. Vind.* 2.127–77.
3 *RMR passim.*
4 E.g., Watson (1956); Syme (1959); Gilliam (1962); Davies (1967); Fink (1971); Bowman (1974); Birley (1994); Bowman & Thomas (1984, 1991); Wilkes (2001); Phang (2007).
5 Vegetius, *Mil.* 2.19f.
6 Cf. Erdkamp (2002) 6.

century to Marius in the late second century, whereas the Marian reforms receive a whole chapter to themselves.[7] This brevity of treatment highlights a gap which requires further research. With regard to military administration, Wilkes goes so far as to state that the 'key stages [...] are evident in the long reign of [...] Augustus [...], notably in the creation of a central resource of statistics and information based on the compilation of written documents forwarded from all over the empire to a central bureau in the capital'.[8] Such an interpretation is untenable. Wilkes suggests that Rome went almost overnight from a position of short-lived, seasonal campaigns with no administration in the Republic to a complex, centralised system governing the long-standing, far-distant legions of the Principate. The record of Rome's military achievements in the Republic, growing from an Italian *polis* into the dominant Mediterranean power, is implausible without some type of administrative backing.

Indeed, the problems to be expected from a poorly administered army operating at length away from its hub – such as mutinies over a failure to arrange payment, food supply or overly long service – are almost non-existent in the history of the Early and Middle Republic. This is all the more significant because such risings are precisely the type of unusual event which ordinarily attracted the attention of ancient writers. The overall lack of mutinies in the extant material suggests that they were not occurring. Even allowing for the gaps in the ancient sources, the mutiny at Italica in 206 stands out as an irregularity.[9] Moreover, the relative ease with which Scipio was able to end the mutiny (by paying his men) and the lack of other Mid-Republican examples indicate an administrative system which ordinarily functioned effectively. The later emphasis on limited terms of continuous service reflects an administration well aware of potential problems and bureaucratically organised enough to avoid them.[10] Mutiny became more common only within the context of the collapsing political system in the Late Republic.[11] This indicates that the existing mechanisms failed to withstand these new pressures; Augustus' administrative 'reforms' could be viewed as a reinstatement of lapsed organisation.

Other modern scholars provide a more balanced assessment than Wilkes regarding the origins of military administration. Most recently, in his discussion of military logistics, Roth argues that it was the development of central (in Rome) and local (in the legion) administration which allowed armies to operate away from Rome for extended periods. He dates this

---

7 Keppie (1984) 14–56.
8 Wilkes (2001) 32.
9 Livy 28.24.5–25.15; Appian, *Hisp.* 34–6.
10 See I:iv.
11 Messer's catalogue of mutinies counts more than 30 from the Jurgurthine War to the end of the Republic, Messer (1920) 170–3.

development to the third century, when Rome began to engage in longer and more distant wars.[12] However, as his study focuses on logistical considerations, he does not suggest the possible nature of these records. Nonetheless, Roth highlights that the historical record as it stands indicates the necessity of administration supporting the successful operation of the legions, emphasising the need for further study.

One suggestion concerning the nature of Republican administration has been made. In the introduction to his catalogue of military papyri from Dura-Europos, Fink acknowledges that this administration's development must have begun during the Republic. He argues that the census was the beginning of military administration – the purpose of its creation was as a record of military manpower[13] – and notes the similarity of census declarations to early military records from the Principate. He suggests that for the legions to operate, a legion roster containing the different lines of *velites*, *hastati*, *principes* and *triarii* must have been necessary and carried with the army, before being discarded when the legion was demobilised at the end of a campaign. Fink suggests that these rosters developed into a more complicated record system sometime in the second century, as armies spent longer and longer in the field. He does not provide any evidence for these suggestions beyond reasonable conjecture,[14] but, with Roth, he thus demonstrates that the origins and development of military administration are areas which require further detailed study.

Outline of approach

The aim of this monograph is to contribute to the study of Roman military administration by demonstrating that the operation of military administration and its development are visible in some detail prior to Augustus. In order to do this, it examines – independent from later evidence – what administrative tools were available in the Republic, how they were implemented and who was responsible for their management. This bureaucratic development can be particularly noted through the Middle Republic, when small, local campaigns gave way to extended wars overseas. In particular, it was forced onwards more quickly by the new scale and scope of military operations brought about by the pressures of the Hannibalic War. Many of the features of the military bureaucracy, both before and after the war, are revealed in the changes which took place during or in the aftermath of the Second Punic War. The monograph therefore addresses several areas which have received little detailed attention in the past, due to a perceived lack of evidence. The discussion remedies the lack, demonstrating

12 Roth (1999) 244–5.
13 Livy 1.44.
14 Fink (1971) 6–7.

that a careful examination of the ancient literary sources yields a great deal of material concerning military administration despite those sources rarely addressing the subject directly. Combining this evidence with archaeological and topographical material and a less traditional, more scientific, methodologically rigorous approach, where appropriate, allows a clearer interpretation.

For the purposes of this discussion, the Middle Republic is defined roughly as 338–146.[15] As well as encompassing the crucial evidence from the Hannibalic War, this limit is imposed in large part by the source material. Polybius and Livy provide most of the evidence, often in the form of passing references to features with which they assume their audiences are reasonably familiar. In combination, these two authors provide a continuous narrative for the vast majority of the period. In addition, an important reason for ending the period in the mid-second century is to avoid the issues brought up by discussion of the Gracchi, the 'reforms of Marius' and the growth of warlords. While the military repercussions of these events are certainly worthy of discussion, they represent a later stage of military development. The aim here is to examine the earlier stages of military documentation in order to understand its mechanisms and methods. This requires a discussion of the army before it reached its 'professional', largely uniform state sometime in the Late Republic. The army of the Principate was the result of extended development over several centuries; it follows that the administration which supported it developed likewise.

The Mid-Republican army under discussion here was for the most part the manipular army described by Polybius.[16] It was composed of three lines of heavy infantry: the *hastati*, the *principes* (both in maniples of 120 men) and the more experienced *triarii* (in maniples of 60) in reserve. They were supported by the light infantry (the *velites*) and light cavalry. The tactics employed by this legion are described by Livy and need not be repeated here.[17] From the Spanish campaigns in the Hannibalic War onwards, both Polybius and Livy refer to the use of cohorts, a unit of three maniples and a complement of *velites*.[18] Scholarly opinion is agreed that these mentions are not anachronistic; rather, they demonstrate the development of a new tactical form to meet situations demanding more manoeuvrability than the

---

15 This corresponds roughly with Flower's first and second 'Republics of the *nobiles*', Flower (2010) 33. While this an interesting approach, this author does not entirely agree with those divisions and they will not be used here.

16 Polybius 6.21.6–23.16; cf. Gilliver (1999); Dobson (2008) 47–8. For a variant on the traditional interpretation, see Armstrong (2020).

17 Livy 8.8.3–14; cf. e.g. Keppie (1984) 33–5; Oakley (1998) 451–66; Gilliver (1999) 15f; Potter (2004) 67–73; Lendon (2005) 178f; Koon (2010) 24; Taylor (2014); Le Bohec (2015) 116.

18 Livy 25.39.1 (210); Polybius 11.23.1, 11.33.1 (206).

traditional manipular formation.[19] The evidence suggests that the main organisational form of the legion remained manipular throughout the period, with the cohort used as necessary. One of the final extant mentions of the manipular legion is given by Sallust, in the context of the Jugurthine War. He demonstrates the ease with which the soldiers could change their formation from one to the other.[20] Sallust is doubly useful here, demonstrating that the switch from maniples to cohorts could be made on an ad hoc basis as well as indicating that the cohortal legion was a later development than the period covered by this discussion.[21]

Sources of evidence

Despite attempting to involve more scientific elements, the discussion is heavily reliant on the literary sources. Thus it is worth briefly highlighting the merits and problems inherent in dealing with this material. Polybius is a key ancient source for this study.[22] His work is particularly valuable for several reasons. Firstly, as a contemporary with the latter end of the period and an experienced military man, his account is transmitted through the lens of an individual familiar with the political and military realities of the time. Further, his close relationship with several leading Roman families, notably the Aemilii and Scipiones, provided him with a perspective from the heart of Rome.[23] Secondly, Polybius' stated intention was to write for an audience which he considered unfamiliar with Roman institutions and practices.[24] This makes his account even more valuable, as he describes features which Roman authors did not. The military digression in book 6, which details the recruitment process from beginning to end, the organisation of the legions and of the camp is extremely important for understanding the operation and mechanisms of military administration in the Middle Republic.[25]

On the other hand, by his own admission, Polybius' descriptions do not always cover the full complexity of the institutions described.[26] This can be

---

19 Gilliver (1999) 18–22; Dobson (2008) 59 following Bell (1965) 415f.
20 Sallust, *Iug.* 51.3 with 49.6.
21 The allied contingent of the Roman army, which often doubled the number of men, will not be discussed. It appears that Rome relied on local allied systems which operated independently of Rome's. Polybius (6.20.4–5) states that the allies were instructed how many to recruit and where to assemble with their own commander and paymaster; Livy's numerous references to senatorial demands for allies accord with this, e.g., Livy 21.17.3, 22.57.11, 32.8.2, 6–7, 33.26.4, 40.1.5, 41.5.4.
22 For a more detailed discussion, with bibliography, of the value of Polybius as a military historian see I:i.
23 Polybius 31.24.
24 Polybius 6.2.3.
25 Polybius 6.19–42.
26 Polybius 6.11.3–6.

seen in his failure to mention the cohort in his description of the army. This may be a result of his personal experience. In 151/0 Polybius accompanied Scipio to Spain – where the cohort was most used in the Middle Republic – but that year's campaign was dominated by siege warfare and thus provided little demonstration of the cohort's tactical value. Coupled with Polybius' Greek military background, this led him to underestimate the cohort and thus not describe its use.[27] Moreover, book 6 was probably written before his visit to Spain.[28] His omission of the cohort highlights that Polybius was not infallible. Nonetheless, the descriptions provide more detail than could otherwise be extracted from the extant ancient material.

Finally, Polybius' approach to writing history adds to his reliability as a source. Unlike many native Roman historians, including Livy, Polybius was concerned with validating the information he recorded. He identified three key areas of historical endeavour: personal experience, questioning eyewitnesses and the study and collation of written works. The first two took precedence over the third.[29] Polybius travelled to become personally acquainted with locations, using the opportunity to interview local witnesses.[30] He also consulted inscriptions rather than relying solely on information in other histories.[31] Overall, he displayed a concern for primary evidence more reminiscent of modern 'scientific' scholarship than the work of his contemporaries. This does not rule out mistakes or misunderstandings, but it does suggest that his work has more historical rigour than might be found elsewhere in the works of ancient authors.

The work of Livy, the other main source for this study, is one of those with less concern for authenticating detail.[32] Livy wrote his history during the Augustan period and reflected the literary preferences of the time. His theme and motivation for writing, as set out in his preface, was moral: he would trace Rome's development in order to provide positive and negative exempla of behaviour and attitudes.[33] As Walsh demonstrates, Livy wished to write truthful history in a worthy literary setting, but he had both more skill with and a preference for well-formed prose over historical research. Nonetheless, even with a didactic purpose, his subject was

---

27 Bell (1965) 414; Appian, *Hisp.* 51–5.
28 Rawson (1971) 13–4; Walbank (1972) 134; Walbank (2002b) 278n.4; Sage (2008) 122.
29 Polybius 12.25e.
30 Polybius 3.59, 12.28a.5–6, 35.6.3–4.
31 E.g., Polybius 3.21–6.
32 For a more detailed discussion of Livy's use of other historians, particularly Valerius Antias, see III:ii. On Livy and his methods, cf., e.g., Walsh (1970); Briscoe (1971); Burck (1971); Walbank (1971); Luce (1977); Miles (1995); Jaeger (1997); Feldherr (1998); Chaplin (2000); Chaplin & Kraus (edd.) (2009), especially Briscoe (2009), Oakley (2009) (updating Oakley (1997)) and Tränkle (2009); Levene (2010).
33 Livy Pr.

still Rome's history.[34] His approach was not the same as Polybius', but it does not render his work unreliable.

However, there is little evidence that Livy ever directly consulted the primary material. For example, he chose to follow another historian over Polybius regarding the number of men Hannibal brought to Italy, despite Polybius having gathered the information from Hannibal's own inscription.[35] Admittedly, the historian in question, L. Cincius Alimentus, claimed to have the information from Hannibal himself,[36] but Livy does not even mention the inscription as an authority with which to verify Alimentus' account. Livy relied on the works of preceding historians. His history reveals that he was widely read, but he seems to have followed one narrative for stretches, only introducing others to add detail or to highlight contradictory accounts.

Nevertheless, the resulting work still has considerable historical value, and Livy's method was not entirely uncritical. He frequently gives the reports of two or more historians, demonstrating his awareness of contradictions in his sources.[37] Often he offers no opinion about which account is the most reliable, but including the divergent material allows readers to consider the problem for themselves. In many ways, this is more useful to the modern historian than Livy giving his judgement without mentioning any difficulties. It reveals that the tradition was varied and emphasises the care required when considering the evidence. Moreover, these fragments give an insight, albeit brief, into the works which Livy used, revealing not only his version of events but those in works now lost. Even when it is impossible to come to a judgement due to a lack of other evidence, knowledge that the problem exists is more valuable than if Livy had simply brushed over it.

The occasions on which Livy does not give variant traditions demonstrate that he was capable of reading with a critical eye. The differences in the parallel accounts of Polybius and Livy demonstrate that it is unlikely that Livy's sources were in full agreement for the majority of the period covered by his extant books. Therefore, Livy made decisions about what to include in his work and what to omit. This may have been done with more of an eye to literary than historical needs, but nonetheless it demonstrates discrimination. Moreover, while Livy's purpose may have been moral, the annalistic framework of the majority of his work, taking events year by year, is historical. For his exempla to have their greatest effect, they required context. As a result, although there are weaknesses in Livy's

---

34 Walsh (1970) 287; Oakley (1997) 3, 114–17.
35 Polybius 3.56; Livy 21.38.2–5. A notable exception to this is his visit to the tomb of Scipio Africanus, Livy 39.56.3. Cf. Luce (1977) 101.
36 *FRHist* 2 F5,cf. *FRHist* III:53–5.
37 E.g., Livy 22.36.1–5.

approach from a modern (and Polybian) historical perspective, his work remains a useful source, not least as the only unbroken narrative for 218–167. It is a literary whole, not just an amalgam of earlier material. Like any ancient source, using the evidence presented is not straightforward, but Livy's work has a significant historical value.

Additional evidence is found scattered through the works of later authors. With the exception of Cicero and Varro, all these authors wrote during the Principate. However, despite their greater distance from the period under discussion, their works give an insight into events not found in Polybius and Livy's works. The later sources provide access to earlier, no longer extant material, both directly and indirectly; like Livy, they relied on earlier works to provide their material. For example, Appian cites as a source for his *Spanish Wars* Rutilius Rufus, a contemporary of some of the events narrated.[38] Further, Appian does not seem to have used either Livy or Polybius as his source for the *Hannibalic War*.[39] In combination, these two factors suggest that Appian's accounts have some independence from those of Livy and Polybius. Therefore, his work can be used to supplement and reinforce (or undermine) the narratives provided by the other two authors. His sources, though, were probably those also used by Livy, and thus his work is not entirely independent.[40] Nevertheless, Livy and Appian used their material differently, so Appian's work can be used as a balance to the former's account.

Imperial sources must be used with care. It is not always clear that the authors understood the world they described, or understood precisely what their sources meant. The world of the second century AD was far removed from the second century BC. As Richardson states well, 'Appian writes with an intelligent and thoughtful appreciation of the problems of empire, but from a standpoint which belongs to his own time, rather than that about which he writes'.[41] All historians, ancient and modern, are influenced by their experience of their own time, however careful their approach. When applying the works of Imperial writers to the Middle Republic, this must be kept in mind. However, as with Livy, the works can still provide useful supplementary information concerning military administration.

The works of so-called antiquarians, or those not writing narrative histories, are a different case. Aulus Gellius' *Attic Nights* provides a good example. Gellius was concerned with gathering assorted material to match his themes, leading him to quote directly from different sources. These verbatim transcriptions, ranging from Piso to the military oath of obedience, provide direct access to Republican material which is otherwise lost.

---

38 Appian, *Hisp.* 88.382. Cf. *FRHist I* 278–81.
39 Richardson (2000) 4–5.
40 Richardson (2000) 4–5.
41 Richardson (2000) 7.

It is reasonable to assume that Gellius correctly copied these passages, as he often quoted to demonstrate an unusual point of grammar or form of words.[42] Careless transcription would defeat his object. The practice of direct quotation means that any problems concerning Gellius' understanding of the unfamiliar Middle Republic are circumvented. Use of the material is not reliant on the interpretation of a later intermediary. In the case of Piso, the problem of a historian's understanding is still present, but the subject is now a Republican familiar with Republican institutions and their operation, even if not always a direct contemporary to events described. In the case of the military oath, even this is not a problem, as it is not a description but the object itself.

In addition to the literary evidence, physical evidence will be used to supplement and strengthen the argument. Topographical and demographic methods will be especially prevalent, in order to address the more practical elements raised by the discussion of the literary evidence. In particular, topography and demography will be used in combination to approach the question of where the recruitment levy, the *dilectus*, was held (I:iv). This approach has not been used before to the author's knowledge; it provides the basis for a more solid discussion of the literary evidence than the speculation to which it was limited previously. Topography will also be used to propose the storage location of the census records (VI:iii), and demography for the operation of the census itself (II:ii).

However, there are limitations to these approaches. These are largely negated by using the physical evidence in combination with the literary, but it is worth outlining them here. They will be discussed in more detail as they arise. For topography, the limits are mostly imposed by the state of the archaeology. The nature of the Mid-Republican *forum* – that is, the size and placement of buildings within it – is not fully known. Many of the sites have been rebuilt upon at a later stage, obscuring the view of what was there beforehand. As well as natural redevelopment as need changed, these problems have also been caused by fire damage and Augustus' rebuilding programme.[43] The major excavation of the *forum romanum* ceased at the Augustan level; despite smaller-scale deeper excavations, much evidence of the Middle Republic remains hidden. In addition, ancient notices of rebuilding rarely describe a location. As will be seen with the discussion of the *aerarium Saturni* and the *atrium Libertatis*, ancient authors assumed that their readers would be familiar with these buildings (V:iii).[44]

---

42 Baldwin (1975) 86–7; Holford-Strevens (1988) 53, 56; Stevenson (2004) 122, 134–6; *FRHist I* 70; cf. Gellius, *NA* pr. 2.

43 *Res Gestae* 19–21.

44 Cf. Pearson (2018) 559–65 on the topographical history of the *forum*.

The major limitation of the demographic approach is that it is based on a model, the Male West Princeton model life table.[45] This is a model accepted for use for ancient Rome,[46] but it was specifically designed for populations about which very little demographic data are known. This study applies the model to the census figures (with alterations as discussed at 1:iv). However, as the model is just that – a model – it can only provide an idea of what a plausible demographic situation may have been. It does not provide a definitive answer. This is reflected here by the use of three different models based on Male West in an effort to cover the plausible extremes of the demographic profile of the male citizen population. The models lend a sense of scale to events described only abstractly by ancient authors. It is this scale which allows more detailed conclusions to be drawn about the issues under discussion.

Chapter outline

The monograph reveals that the outline of military administration conjectured by Fink is supported by the evidence, but much more additional detail can be uncovered by close examination. Despite the lack of extant documentation, a great deal can be understood concerning the paperwork which enumerated Roman manpower and tracked it on campaign. This includes the mechanisms in place in Rome and in the legions themselves as well as those responsible for maintaining them. Fink was wrong to imply that the failure of any Mid-Republican documentation to survive renders the administration irrecoverable. Moreover, the development of this bureaucracy in direct response to changes in the scope and scale of Rome's wars will be seen, demonstrating the ability and willingness of the state to react to external pressures in order to operate most effectively.

In order to examine the military administration of the Middle Republic, it is first necessary to establish what the documents themselves were. The logical place to begin the discussion is with the recruitment of soldiers, the first step in organising the legions. Chapter I provides a detailed examination of Polybius' description of the levy, taking each passage separately. Each of the issues raised – such as location, soldier selection and legion size – is examined individually. Investigating each of these elements reveals the requirement for and existence of lists of the liable and the legions there created, as well as demonstrating Rome's flexibility in meeting military challenges appropriately.

Chapter II examines the place of centralised administration in military bureaucracy. The need for lists of those liable for military service is established in the first chapter; the census is the place from which such a list would originate, indicating its role in military administration. In order to better

---

45 Coale & Demeny (1983) 107, 110.
46 Parkin (1992) 80; Saller (1994) 25. See I:iv.

understand its role, census declarations are examined. The registration of those on campaign is discussed, revealing that the census operation was remarkably adept at tracking all male citizens even when they were unable to attend the censors. Finally, the manpower figures for an emergency levy in 225 are investigated, demonstrating that despite the complexity of the system revealed by the foregoing discussion, Rome was able to circumvent these complications when the threat made it expedient. Problems of 'red tape' could be avoided to allow an effective, organised response before incorporating these extraordinary events back into the records in order to prevent citizens feeling unduly put upon by military service.

The next area of documentation examined is that generated on campaign. This area is more difficult to investigate but is nonetheless vital for a fuller understanding of the bureaucratic mechanisms operating around the Roman army. Chapter III focuses on three areas: enumeration of the dead, pay accounting and tactical strength. All three emphasise the importance of legion lists generated at the levy to running the army in the field. The first demonstrates that legion commanders were able to identify battle losses, those either killed or missing, in some detail. It is probable that these lists could be, and were, sent to Rome with dispatches, keeping the city abreast of the situation in the field. Discussion of the payroll reveals that a separate record was required for each soldier, as deductions made for items such as replacement equipment were not consistent across even those of the same battle line. Thirdly, examining the reinforcements, *supplementa*, sent out to forces in the field reveals that Rome also had a reasonable idea of the real strength and disposition of its legions. This was coupled with an awareness in Rome that delays in communication could have resulted in more losses yet to be reported, a problem for which the recruiters endeavoured to compensate. A review of the forces carried out by a new commander ensured that any errors which had arisen in the legion's accounting were rectified.

Chapter IV moves back to Roman soil in order to examine the administration and processes behind the financial support for Rome's wars. It discusses the collection of Rome's only direct tax, *tributum*, and its distribution to soldiers as *stipendium*. Direct evidence is scant, but close examination reveals a fundamental change in practice during the third century. Changes in the scale and scope of Rome's wars centralised what had previously been a localised system of military pay organised by tribal officials entirely independent of the army or central state. This change increased the responsibilities of quaestors both in Rome and on campaign. More importantly for this monograph, it necessitated the development of the pay regime described by Polybius and its attendant paperwork.

The first four chapters combine to create a complex picture of Rome's bureaucracy. They show a state fully aware of the benefits of accurate enumeration and working to achieve them. However, a full discussion of military administration in the Middle Republic must also take into account more practical considerations. The physical nature of documentation would have had a profound effect on its use: large, heavy documents would have had a distinct tactical disadvantage to the legion needing to transport them. On the other hand, documents needed to be sturdy enough to survive the rigours of a campaign. In Rome, records such as the census did not suffer such rough treatment, but they too needed to survive well until at least the next census period. The storage locations of such documents is also a consideration. The possible volume of documents has an impact on positing buildings, which themselves are not easily identified. Chapter V addresses these issues, demonstrating that these limitations are not an impediment to the picture of bureaucracy developed in the preceding chapters.

Finally, Chapter VI discusses those responsible for creating and managing records. A major potential criticism of this monograph is that Rome was not yet sufficiently literate in the Middle Republic to support such a complex system. To some extent this is countered by the existence of this bureaucracy itself: if such an administration was impossible, it should not be visible in the source material. This chapter addresses the issue of literacy more closely, demonstrating that there is evidence of widespread semi-literacy by the third century, sufficient for the needs of the majority of the soldiers. More complex work was limited to the better-educated, higher social classes, but such an education was expected of them, not an exception. Following this, the individuals responsible in Rome and on campaign for particular parts of the military administration proposed are examined. These elements will complete the picture of a Rome both willing and able to develop its bureaucracy in order to make its military operations as effective as possible.

It is only once this examination has been completed that the findings will be compared with the extant documentation. The comparison in the Conclusion reveals perhaps a surprising level of continuity, given the changes to the nature of the army between the third century BC and the second century AD. This demonstrates that Augustus did not invent Roman military bureaucracy but rather reestablished the tried and tested methods of the Middle Republic.

1 Dilectus

When examining the bureaucratic record of Republican military service, the obvious place to start is at the beginning of that service: the *dilectus* (recruitment levy). The longest and best ancient description of the *dilectus* is provided by Polybius, where it begins his military digression in book 6 of the *Histories*[1]. Here he recounts the process for selecting military tribunes and legionaries, and how they are divided into legions. He gives the maximum terms of military service alongside the minimum required for political office. An oath is administered and a day set for the men to reconvene. The organisation of the legion into battle lines and the selection of unit officers is described. Finally, the men are instructed to arm and given the day on which to attend the consul. This chapter will examine Polybius' levy in order to demonstrate the necessity of written administration for the process to function. Beginning with a detailed examination, the chapter's focus then broadens to include aspects apparently missing from Polybius' account. This creates a full picture of the Mid-Republican *dilectus* and of administration's integral role within it.

i Polybius as a military historian

First, however, the authority of Polybius' account must be established. Scholars generally accept that book 6 was published in the late 160s or early 150s,[2] while Polybius was one of a thousand Greek hostages held in Italy. Of the thousand, Polybius alone was held in Rome.[3] He probably stayed with a senatorial family and had a close relationship with the Aemilii and Scipiones through P. Cornelius Scipio Aemilianus, providing excellent access to those who ran the empire.[4]

---

1 Polybius 6.19–26.
2 Rawson (1971) 13–14; Walbank (1972) 134, (2002b) 278n.4; Sage (2008) 122.
3 Polybius 30.13; Pausanias 7.10.11; Livy 45.31.9. Cf. Erskine (2012) 17.
4 Polybius 31.23–4.

Despite this, some have judged the account anachronistic for Polybius' own time and even the late third century. Walbank argues that it was largely derived from Polybius' personal observations and enquiries concerning the army, but warned that its location in the narrative must be taken into account.[5] Book 6 is immediately after the disastrous Battle of Cannae in 216. Inserting the digression here demonstrated the rigidity and strength of Roman institutions, which were, for Polybius, the determining factors in Rome's coming dominance.[6] He also placed great emphasis on what he considered proper modes of historical enquiry (personal experience, eyewitness questioning and the study and collation of written works[7]), castigating those writers who failed to meet his standards.[8] Taken together, it seems that a conclusion of reliability, for the late third century at the very least, can be safely reached.

However, a writer's ideals do not always translate into reality, and Polybius is guilty of this elsewhere in his work.[9] Modern scholars repeatedly condemn the military digression as outdated and occasionally implausible, particularly in placing the levy on the Capitol.[10] Rawson suggests that the digression's place should not intrude on how Polybius' understanding of the *dilectus* process is interpreted; she allows that he believed he was describing the current system.[11] Rather, she sees the account as originating – and most likely directly lifted – from an old handbook for military tribunes, dating to c. 210 at the earliest.[12] Her suggestion has been followed by a number of other scholars.[13]

There are problems with this solution. Firstly, it is unclear why Polybius (who usually placed great weight on gathering accurate information) would accept this outdated account without further enquiry. He was in Rome for several years, meaning he should have noticed whether Rome's manpower descended on the Capitol each year. (Indeed, if he was in Rome but did not realise it was occurring, the assumption that it caused such great disruption

5   Walbank (1957) 698–9, (2002b) 279.
6   Polybius 6.2; Eckstein (1997) 175f; Brink & Walbank (1954) 115f; Pelling (2007) 247.
7   Polybius 12.25e.
8   Polybius passim, but especially 12.
9   E.g., he criticises Timaeus' use of only records but proudly recalls inscriptions he has used, Polybius 12.11.1–2, 3.22; speeches are also reproduced, Walbank (1972) 43–5.
10 Taylor (1957) 342 n.15; Momigliano (1975) 25; Rawson (1971) 13 (although she does not find the levy on the Capitol implausible); Brunt (1971) 627; De Ligt (2007) 115–16; Dobson (2008) discussing the camps at Numantia; Erskine (2013) 238. The probability of the *dilectus* occurring on the Capitol both before the Hannibalic War and during the period Polybius was in Rome will be discussed below (I:iv).
11 Rawson (1971) 13.
12 Rawson (1971) 15. However, she does conclude that it is unlikely that Polybius' account was entirely wrong, and considers a levy in Rome plausible.
13 Dobson (2008) 54; Sage (2008) 122.

as to render it impossible must be questioned.[14]) Secondly, if the account is to be considered outdated for the late third century, as Brunt does,[15] why was it still being recorded during the Hannibalic War? Finally, the emphasis on the military tribune over the consul is hardly unique. With six tribunes per legion and only one consul per army (often of two legions and equivalent allies), tribunes necessarily featured prominently in running and leading a legion.[16] It is not surprising that military tribunes therefore also feature prominently in the literature. Rather than impugn the comprehensiveness of Polybius' account, the tribune's prominence should be seen as a reflection of everyday practice in the legions. The inconsistencies in the arguments marshalled against accepting Polybius' account of the *dilectus* mean a different interpretation is required.

Polybius was above all a military historian, lacking the temperament for more traditional scholarly interests such as Homer and philosophy, preferring instead warfare's technical details.[17] This does not mean that Polybius was unfamiliar with or incapable of producing such works; his discussion of the Roman constitution proves the opposite.[18] Rather, the military sphere was simply of greater interest to him. Polybius himself was keen to emphasise the difference between his own methods and those of his more 'literary' counterparts. Polybius was not simply an observer, but had a military reputation in his own right; he provides an 'educated officer's view'.[19] He served as a cavalry commander in the Macedonian War, a position which Walbank points out often led to the generalship of the Achaean League.[20] Walbank further highlights several achievements which point to a successful career in Achaea.[21] This reputation is supported by consul M'. Manilius ordering Polybius to report to Lilybaeum for the invasion of Carthage in 149.[22] Many aspects of ancient warfare were universally applicable, but it is unlikely that Polybius was summoned to provide tactical advice for an army about which he had little or no knowledge.

More importantly, the success of Polybius' work highlights his understanding. The work was addressed to a Greek audience. Providing an account of the *dilectus* which he knew to be incorrect or outdated would have

---

14 Brunt (1971) 625.
15 Brunt (1971) 627–8.
16 Polybius 6.19–42; see VI:ii.
17 Walbank (1972) 33; Marsden (1974) *passim*; McGing (2010) 175.
18 Polybius 6.1–18, 6.43–59. Cf. Champion (2004) 87.
19 Armstrong & Fonda (2020) 1.
20 Polybius 10.22.9; Walbank (2002a) 21.
21 Walbank (2002a) 20–21: Polybius was chosen to carry the ashes at Philopoemen's funeral (Plutarch, *Phil.* 2.15), sent in embassy to Q. Marcius Philippus in 169 (Polybius 28.6.9) and requested by Ptolemies VI and VIII to lead the troops sent to them (Polybius 29.23.7).
22 Polybius 36.11.1.

undermined Polybius' stated purpose of helping Greeks understand their conquerors.[23] The Roman elite were also familiar with the *Histories*. Cato the Elder ridiculed Polybius for painting himself as a modern-day Odysseus in his attempts to visit distant places.[24] Cato notably fails to make any (recorded) comment concerning the historical or military content of Polybius' work, suggesting that nothing was considered glaringly incorrect. Too much weight should not be placed on this point, as it is impossible to know what has not survived. Nonetheless, it appears that Polybius had a justified reputation in military matters which his *Histories* only enhanced.

Finally, the text itself reveals an awareness similar to his reputation. The notes on changing practice in cavalry enrolment and weaponry developments show that if Polybius was following an earlier source, he was not doing so blindly.[25] Indeed, the section on weaponry change reads very much like an addition; the sentence structure becomes much more convoluted. Significantly, the note on legion size can be removed from the text without damaging the narrative's flow or sense.[26] Thought, selection and addition are visible in the construction of the *dilectus* account. This accords with the image of a writer very much in command of his subject matter.

ii Terminology

Before examining Polybius' text in detail, it is worth highlighting the language used by both Greek and Latin writers to describe the levy. This opens a window onto their perceptions of the levy and demonstrates the integral nature of bureaucracy in the *dilectus*.

First, *dilectus* itself.[27] The noun is derived from the perfect participle of *diligo*, a verb meaning 'to distinguish by selecting from others'.[28] This in itself does not necessarily imply a written record. However, closer examination of the verb's origin is helpful here. *Diligo* might initially be considered to have been created from *dis-* and *ligo*. However, the form from which *dilectus* originates, as indicated by the perfect participle's form, is *dis-lego*. The first principal part, *diligo*, appears unusual, but has undergone a vowel shift that is to be expected with this combination of consonants.[29] This origin gives *dilectus* a sense of 'to select by reading out', implying the

23 Polybius 6.2.3; Momigliano (1977) 71; Erskine (2013) 231; Thornton (2013) 213–15.
24 Polybius 35.6.3–4.
25 Polybius 6.20.9, 6.25.6–11.
26 Polybius 6.20.8.
27 E.g., Livy 21.26.2, 22.2.1, 26.31.11, 27.38.1, 29.13.1, 32.9.1, 32.26.12, 35.2.8, 38.44.8, 39.20.4, 40.1.3, 41.5.4, 43.11.10, 44.21.5; [Livy], *Periochae* 14.3, 55.2.
28 *OLD* s.v. *deligo*[1] (i.e., *diligo*. The *OLD* s.v. *diligo*[2] is mistaken when referring to *deligo*[2]).
29 Philomen Probert (pers. comm.), David Langslow (pers. comm.). *OLD* s.v. *deligo*[1] gives *de-lego*, a later stage of development.

presence of some sort of list. This does not give any indication of the degree of bureaucracy involved, but it may indicate a process that was at least partly written.

The levy is consistently described in terms which refer to a written process. In Latin, *scribere*, 'to write', is extremely common.[30] Unlike *dilectus*, which requires only reading, *scribere* indicates that writing was key to the levy. The other word frequently used to describe the levy is *conscribere*, 'to write down together'.[31] It is difficult to interpret this as anything other than the creation of a new list. Polybius uses a similar Greek term, καταγάφειν.[32] Again, this has a literal meaning of 'to write down'. It could be a literal translation of *conscribere*,[33] or it could be used for its similar Greek meaning of 'to list, register or mark out'.[34] On either interpretation, its usage supports the creation of a new legion list during the levy. This does not rule out making a mark of some type on a census record or something similar to indicate service, but it also does not confirm it. Importantly, the consistent use of *scribere*, *conscribere* and καταγάφειν for the levy suggests that written processes were integral to it.

This chapter is limited to discussing Polybius 6.19–20 because of the language Polybius uses. After this portion of the description (the selection of men, their division into legions and mention of legion sizes), he declares the levy complete.[35] Thus the rest of his description (dealing with the first and second reassembly of the legions for division into lines and mobilisation, respectively) is of internal organisation and administration, not part of the *dilectus*. Polybius here uses the phrase ἐπιτελεσθείσης δὲ τῆς καταγραφῆς τὸν προειρημένον τρόπον, 'the list having been completed in the previously stated manner'.[36] This implies the primacy of written administration in ancient writers' very conception of the *dilectus*.

iii Service terms

Polybius begins his military digression with a description of service terms. This includes both the maxima which could be served by foot and horse and the minima required for public office in the army and civilian life:

---

30 *OLD* s.v. *scribo*; e.g., Livy 21.17.3, 22.11.2,3, 23.24.5, 24.11.6, 25.3.4, 26.1.12, 27.22.6, 29.13.1, 30.2.1,6, 31.8.5,11, 32.1.3, 33.25.10, 34.56.4, 35.20.4, 36.1.6, 37.2.4, 8–10, 38.35.9, 39.20.3, 40.1.5, 41.9.2, 42.18.6, 43.12.5–6, 44.21.5.

31 *OLD* s.v. *conscribo*; e.g., Livy 21.26.2, 22.11.8, 24.20.1,13, 27.46.3, 28.10.14, 30.41.5, 36.3.13, 37.2.6, 40.1.5, 40.26.5, 40.28.10, 41.5.4, 41.21.5, 43.11.10, 44.21.8.

32 Polybius 6.19.5 ποιεῖσθαι τὴν καταγραφὴν, 6.21.1 τῆς καταγραφῆς.

33 David Langslow (pers. comm.).

34 LSJ s.v. καταγραφή.

35 Polybius 6.21.1.

36 Polybius 6.21.1.

ἐπειδὰν ἀποδείξωσι τοὺς ὑπάτους, μετὰ ταῦτα χιλιάρχους καθιστᾶσι, τεττρασκαίδεκα μὲν ἐκ τῶν πέντ᾽ ἐνιαυσίους ἐχόντων ἤδη στρατείας, δέκα δ᾽ ἄλλους σὺν τούτοις ἐκ τῶν δέκα. τῶν λοιπῶν τοὺς μὲν ἱππεῖς δέκα, τοὺς δὲ πεζοὺς ἓξ <καὶ δέκα> δεῖ στρατείας τελεῖν κατ᾽ ἀνάγκην ἐν τοῖς τετταράκοντα καὶ ἓξ ἔτεσιν ἀπὸ γενεᾶς πλὴν τῶν ὑπὸ τὰς τετρακοσίας δραχμὰς τετιμημένων· τούτους δὲ παριᾶσι πάντας εἰς τὴν ναυτικὴν χρείαν. ἐὰν δὲ ποτε κατεπείγῃ τὰ τῆς περιστάσεως, ὀφείλουσιν οἱ πεζοὶ στρατεύειν εἴκοσι στρατείας ἐνιαυσίους. πολιτικὴν δὲ λαβεῖν ἀρχὴν οὐκ ἔξεστιν οὐδενὶ πρότερον, ἐὰν μὴ δέκα στρατείας ἐνιαυσίους ᾖ τετελεκώς.[37]

(Polybius 6.19.1–4)

Whenever they elect consuls, after this they appoint military tribunes, fourteen from those already having five campaign years, and ten others from those having ten. Of the remaining, it is necessary that the horse complete ten campaigns, the foot six <and ten> by compulsion within forty and six years from birth except those valued below 400 drachmae; they let all these fall into naval service. If at any time some emergency should oppress them, the foot are obliged to serve twenty campaign years. It is not permitted for anyone to take political office before he has completed ten campaign years.

Unfortunately, part of this text is corrupt. Where modern editions supply ἓξ <καὶ δέκα> δεῖ as quoted here, the manuscripts give either ἓξ οὐ δεῖ, '… six; it is not necessary…', or ἓξ οὕς δεῖ, '… six. It is necessary for them …'.[38] Moore argues that the two oldest manuscripts containing this section of book 6 must derive from a single exemplar, as they contain exactly the same excerpts.[39] However, the negative provided by these two manuscripts (οὐ) does not make sense. The relative οὕς is a more plausible reading (making the ordinary service term for the foot six years) but produces difficult syntax. Alternatively, following Cavaignac, οὕς/οὐ can be deleted entirely, again leaving a six-year term.[40]

Modern scholars, as shown in the quoted text, follow Buettner-Wobst in supplying και δέκα.[41] This gives a normal term of 16 years rather than six. A 16-year term is usually accepted by modern scholars because it is also the original service length established by Augustus when he formalised his professional army.[42] However, as Brunt points out, this does not mean that 16 years' service was normal in the second century, nor that the term was

37 Buettner-Wobst's emendation used in the Loeb has been corrected from καὶ <δέκα> to <καὶ δέκα>.
38 FS; D²G.
39 FS; Moore (1965) 55 (stemma on 73).
40 Cavaignac (1914) 76.
41 Buettner-Wobst (1889) 265.
42 Walbank (1957) 698; Brunt (1971) 399; Southern (2007) 92; Nicolet (1980) 97; Harris (1985) 45n.1; Keppie (1984) 33; Dio 54.25.6.

unchanged between Polybius and Augustus.[43] This is a valid caveat, especially as 16 is a modern emendation. This section argues that the ideal normal service length was six years, while the ultimate maximum of 20 years gave the senate enough leeway to cover emergencies and render service extremely flexible. Through the service terms a process of officially recording service length can be glimpsed.

Appian's narrative suggests that six years was the normal length of continuous service in the second century.[44] Keppie and Nicolet have argued that political pressure made six years the ideal (although not legal) limit to continuous service.[45] Consuls were increasingly unwilling to enrol those reluctant to serve;[46] occasional mutinies demonstrate that soldiers would object if they considered their service too long.[47] The terms served before mutiny varied, but it is notable that all (outside the Hannibalic War[48]) occurred within eight years. Brunt argues that by the mid-second century, six years had been the normal service term in Spain for some time.[49] Appian's evidence is for Spanish campaigns, as is the reluctance to serve encountered by the consuls. This six-year term seems to have been a response to pressure exerted by unrest among long-serving Spanish legions. A six-year service term can only be stated with certainty for Spain, although this does not exclude it for the rest of the empire.

Richardson considers Spain a special case.[50] He cites the second-century satirist Lucilius to demonstrate the special treatment required for men serving in Spain.[51] Lucilius states that a soldier served in Spain for thrice six (i.e., 18) years. However, the comment exists only as a fragment, lacking context. It is unclear whether the 18 years were continuous. More importantly, 18 years of service would have been excessive for anyone other than a career soldier, whether or not the individual served in Spain. If the remark was calculated to be inflammatory or striking, 18 years has greater impact in the place of six than 16. Lucilius' 'thrice six' (*ter sex*) fits the metre, but may also serve to

---

43 Brunt (1971) 399.
44 Appian, *Hisp.* 11.61, 11.65, 13.78, 14.86. Cf. Livy 34.56.8, 40.36.7. Only 13.78 makes a direct reference to six service years, but the years between fresh troop recruitment suggests that approximately six service years was the norm during this period.
45 Keppie (1984) 33; Nicolet (1980) 113.
46 Livy 43.14.2.
47 Cf. Messer (1920). It is notable that the mutiny against Scipio during the Hannibalic War was the first safely attested mutiny in which service length can be deemed to have played a major part. It is in the late second and first centuries during the civil wars that mutiny easily linked to long service terms can be most readily observed.
48 Livy 28.24–32; Polybius 11.25–30; Appian, *Hisp.* 34.137–36.146; cf. Chrissanthos (1997).
49 Brunt (1971) 401.
50 Richardson (2000) 167.
51 Lucilius 15.509–510 – *dum miles Hibera terrast atque meret ter sex aetati' quasi annos.*

support the six-year limit.[52] The lack of context for the remark means that the tone cannot be inferred, whatever might be expected of the satirist. There is no sense of judgement in the surviving fragment; it simply states the case. Thus the Lucilius fragment is unhelpful in determining whether Spain was a special case in setting continuous service terms.

Nonetheless, the evidence from Livy and Appian suggests that it was routine to send men home after six years' continuous service. Nicolet suggests that the 184 decree to discharge all those who had served their term in Spain referred to either ten or 16 years.[53] However, Livy gives no such indication. He refers only to those who had completed their service, *emerita stipendia*. Moreover, Spain had been receiving reinforcements over the previous years, with veterans dismissed in the same manner.[54] While the legions remained nominally the same, their composition changed. It is unlikely that anyone dismissed in 184 had begun serving before 193, that is, nine years before. Indeed, the army in Further Spain had been taken out in 195 by M. Porcius Cato.[55] This does not demonstrate that the maximum term was six years, but the reinforcements and dismissals demonstrate a desire to limit continuous service terms. It can be inferred that nine years was considered a long term.

At a tangent to this is continuous versus noncontinuous service. Southern states that Roman men could serve only six years continuously, but remained liable for up to 16 years of service.[56] The evidence examined thus far neither confirms nor denies this. If six years was only the ideal norm for continuous service, 16 becomes a plausible (if conjectural) reading of Polybius; but if six years was the total ideal normal term, modern reconstructions of 16 must be considered anachronistic. The recruitment oath added to the census oath in 169 may help clarify this situation.[57] The oath required all those under 47 who had not served to present themselves to the *dilectus*. This suggests that in 169 anyone who had served at all had fulfilled their obligation.[58] This makes more sense in a context where six years was the ideal norm for service and campaign years were served consecutively. If this was the case, it explains both the unrest of soldiers on a longer campaign and the senate's desire to limit service to this length.

---

52 Tim Cornell (pers. comm.).
53 Nicolet (1980) 113; Livy 39.38.8.
54 Livy 37.50.11 (189), 34.56.8 (193).
55 Livy 33.43; Appian, *Hisp.* 40.161.
56 Southern (2007) 92.
57 Livy 43.14.5–6. To be discussed further in the chapter.
58 Livy 43.14.6 – *si miles factus non eris, in dilectum prodibis?* – 'If you will not have been a soldier, will you be present at the levy?' It is possible to translate the first clause to mean 'if you are not then serving as a soldier', but the author believes that the first translation is a more accurate rendering. However, the problem of translating this phrase is underlined by the English editions which retain the ambiguity.

The extraordinary maximum service term of 20 years can rarely if ever have been reached, even during the Hannibalic War. The survivors of Cannae were decreed to spend the war's remainder in Sicily.[59] The earliest these men were enlisted was 218.[60] According to Livy they were not officially discharged until late 201 or 200,[61] but this is still only a total service of 18 years.

The Hannibalic War was an extreme situation, the duration of which the Romans could not have predicted. The decree which banished the Cannae survivors to Sicily set no limit beyond the war's unknown end.[62] It was intended as a punishment where theoretical limits were to be ignored, but it is unlikely that a war of such duration was imagined even in 216. However, the key point is that no numerical limit was imposed. Effectively limitless service was intended to be a severe punishment, indicating that there was a maximum service term. In the context of 216 this could as easily have been six years as 20. Whichever was imagined, the Cannae veterans demonstrate that serving 20 years was extremely unusual.

The final years served by the Cannae veterans (from the end of the Hannibalic War in 203 to their eventual discharge late 201 or 200) were met with complaints and near mutiny by the men themselves. However, it is described as volunteer service by Livy.[63] The continuing service of the Cannae veterans as reluctant 'volunteers' suggests that if an individual volunteered for a campaign, the number of years served could go beyond the theoretical maximum. This is supported by other examples of volunteering,[64] most notably Spurius Ligustinus' career.[65] Thus, while there was a theoretical military service maximum, it could in reality be circumvented by volunteering (or the appearance of it). Additionally, the *tumultus*, an emergency levy, overrode any exemptions, including *emerita stipendia*.[66] It acted as another method to ignore the service limit. It is therefore possible that Polybius' absolute maximum of 20 campaign years was in fact only a symbolic figure, representing a term ordinarily beyond imagination. This also fits with an ideal normal term of six years rather than 16. The jump from six to 20 is considerably more than from 16 to 20.

59 Livy 23.25.7. It is worth noting that the Cannae survivors who completed their term in 216 were not banished to Sicily. It is highly unlikely anyone had served fourteen years prior to the Hannibalic War given that only four legions were enrolled every year and campaigns tended to be short. Therefore, the fate of the Cannae survivors also suggests that six years was the normal complete service term of citizens, at least before the Hannibalic War.

60 Livy 21.17.2.

61 Livy 30.41.5, 31.8.5.

62 Plutarch, *Marc.* 13.3; Livy 23.25.7 – *in Siciliam eos traducti atque ibi militare donec in Italia bellum esset placuit*.

63 Livy 31.14.2, 32.3.3 (near mutiny), 35.2.8, 37.4.3.

64 Livy 9.10.6, 25.19.13, 27.46.3, 37.4.3.

65 Livy 42.34.

66 See II:iii.

Terms of six years, and 20 in emergencies, allowed the senate to legally mobilise whatever manpower they considered meet whilst simultaneously giving Rome's citizens a more palatable service term.

The service required for holding office is also instructive. It may initially seem odd that senatorial candidates required a longer service record than the ordinary length of infantry service, but there are several possible reasons for this. The ten-year requirement may have been linked to the equestrian service maximum. As the Republic developed, the senate was increasingly limited to those rich enough to qualify for equestrian service (although they remained liable for infantry service if called[67]). The political requirement was the same for all, whether serving as foot or horse. Further, Rome's class system meant that those with a greater stake in the state also had greater responsibilities, as demonstrated by richer citizens being liable for the more expensive cavalry service.[68] Thus it is not surprising that those desiring to have a political (powerful) role in Rome had to provide greater service beforehand.

Moreover, it should not be forgotten that the pinnacle of the Roman political career, the consulship, was a generalship. Harris and Hopkins have pointed out that the majority of a consul's military training came from practical experience gained during ordinary military service.[69] Campbell questions the practicality of this,[70] but Polybius himself provides an answer. The position of military tribune required either five or ten years' service, depending on seniority.[71] The six military tribunes effectively ran the legion day-to-day, and often commanded detachments.[72] On occasion, ex-consuls became military tribunes, demonstrating the position's tactical importance.[73] Further, it is unlikely to be a coincidence that the qualifications for a senior military tribune and for senatorial office were the same: ten years' military service. The military tribunate was often a stepping stone into political office, demonstrating that practical military experience was a suitable education for both an extended military career and a senatorial one.

The extant evidence for office holding in the municipality of Heraclea provides an interesting comparison. The *tabula Heracleensis* provides the required service terms for anyone wishing to serve as a magistrate before they reach 30: three years in the cavalry or six years in the infantry.[74] Using

---

67 Walbank (1957) 700 suggests that the change to enrol the horse before the foot was so that those who were not selected for cavalry service would still be liable for infantry service.

68 Polybius 6.20.9.

69 Harris (1985) 11; Hopkins (1978) 27.

70 Campbell (1987) 20.

71 Polybius 6.19.

72 Polybius 6.19–42, cf. VI:ii.

73 E.g., Livy 42.49.9.

74 *Tabula Heracleensis* II.98–107.

this law is not without difficulty. It is Late Republican in date.[75] It is unclear what terms those above 30 had to serve; they may have been modelled on Roman practice.[76] Unlike in Rome, those serving in the cavalry required less active service than those in the infantry. The service terms are also lower than those required in Rome, probably reflecting the less senior position of local magistrates. The need to legislate for under-30s suggests that Heraclea regularly appointed young magistrates. Nevertheless, despite these differences, the inclusion of service terms allows broad conclusions to be drawn.

Most importantly, military experience was required. Moreover, this experience had to be genuine. The inscription requires that more than six months be spent in camp during the campaign years, or that two campaigns be undertaken in that year.[77] This supports the rebuttal to Campbell that individuals did gain genuine military experience while on campaign. The importance of a military background is further emphasised by the reference to magistrates as *decuriones* and *conscripti*. Both these terms refer to magistrates, but more properly belong in the military sphere. They approximately translate as 'horse captains' and 'the enlisted'. However, *conscripti* can refer to those enrolled in the senate by the censors.[78] It is possible that by the first century the civilian interpretation of *decurio* dominated its meaning. Nonetheless, it suggests that earlier a military role was considered integral to holding office. Even if Heraclea simply adopted Roman terminology, this does not diminish the significance of the term's origin.

All of this has an interesting impact on understanding the legions' written administration. A set service length necessitates recording campaign years served. This would have allowed individuals to argue for an *emerita stipendia* exemption, as well as helping the senate more accurately calculate the liable manpower at its disposal.[79] It should be noted, however, that if any past service was considered equivalent to completed service (as implied by the 169 oath), a detailed record system would not be required. Only a mark against those who had served would be necessary. On the other hand, requiring a specific service length to run for office suggests that more than a yes/no approach was required. The differing lengths for different offices (five years for junior tribunes, ten for senior) also supports this; a yes/no indication would not differentiate.

---

75 Crawford *et al.* (1996) 360–2.

76 Andrew Fear (pers. comm.).

77 *Tabula Heracleensis* II.101–2: *quae stipendia in castreis inue prouincia maiorem partem sui quoiusque anni fecerit, aut bina semestria, qua<e> ei pro singuleis annueis procedere operteat.*

78 Senators are frequently called *patres conscripti*. E.g., Livy 22.60.7; Sallust, *Cat.* 51.1; Cicero, *Verr.* 2.3.82.

79 See II:iv.

That senior tribuneship and senatorial office required more than the six-year normal service term suggests the same. Consequently, it must be concluded that Polybius' service terms demonstrate that a record of years served by each citizen was necessary. This was the case in Rome in at least the late third and second centuries, and remained the case for at least one municipality into the first century.[80]

iv *In Capitolio*

One element in particular causes modern scholars to question the veracity of Polybius' account: conducting the levy on the Capitol, either in Polybius' day or even in the late third century.[81] Polybius explicitly states that the levy is on the Capitol; Livy also refers to it there.[82] From a bureaucratic standpoint, such centralisation is not at odds with Rome's *polis*-like administrative structure. If the *dilectus* operated as described, it would shed light on the use of paperwork in the levy. However, was it really possible to gather so many men to Rome from across the *ager Romanus* and place them on the Capitol? This section examines the problem from a topographical, demographic and literary perspective. It demonstrates both that the *dilectus* could have occurred on the Capitol in the Middle Republic and the primary importance of written documentation in the process.

> ἐὰν δὲ μέλλωσι ποιεῖσθαι τὴν καταγραφὴν τῶν στρατιωτῶν οἱ τὰς ὑπάτους ἔχοντες ἀρχάς, προλέγουσιν ἐν τῷ δήμῳ τὴν ἡμέραν, ἐν ᾗ δεήσει παραγενέσθαι τοὺς ἐν ταῖς ἡλικίαις Ῥωμαίους ἅπαντας. ποιοῦσι δὲ τοῦτο καθ' ἕκαστον ἐνιαυτόν. τῆς δ' ἡμέρας ἐπελθούσης καὶ τῶν στρατευσίμων παραγένων εἰς τὴν Ῥώμην, καὶ μετὰ ταῦθ' ἀθροισθέντων εἰς τὸ Καπετώλιον...

(Polybius 6.19.5–6)

If the consuls in office wish to make a list of the soldiers [i.e., enrol them], they announce beforehand in the assembly the day on which it will be necessary for every Roman in the ages to be present. They do this each year. The day having come and the fit-for-service having arrived in Rome, and after this having been mustered on the Capitol...

---

80 M. Tullius Cicero's career is the first recorded where full military service was not required. His unsuccessful stint as a general may indicate that military experience remained desirable. However, besides the career of the aforementioned Spurius Lingustinus, Cicero's is the first for which a full record survives. It may be that the ten years' service recorded by Polybius was rarely reached in reality, in the first century at least.

81 Walbank (1957) 699; Brunt (1971) 625–6; Sage (2008) 121; Erskine (2013) 238. Nicolet (1980) 100 and Toynbee (1965a) 505 are of the few who accept Polybius' location, although implicitly; Rawson (1971) 15 explicitly accepts it.

82 Polybius 6.19.6; Livy 25.3.8–14, 26.31.11, cf. 39.29.10 (in the city, rather than explicitly on the Capitoline).

Space and capacity

To begin, it is necessary to establish the space available on the Capitol. Examining the tribal assembly, which also met on the Capitoline, MacMullen concludes that c. 25,000 men could fit in the *area Capitolina* if it was unencumbered by statues.[83] However, to the author's knowledge no other attempts have been made to examine how many could fit on the Capitoline Hill. A more detailed consideration is required to establish the Capitoline's capacity, and exactly which areas of the hill were used.

The first difficulty in this task is determining what exactly is meant by the terms *in Captiolio* and εἰς τὸ Καπετώλιον. The Greek appears to be a translation of the Latin term in the accusative. At its most precise, *Capitolium* refers only to the southern elevation containing the precinct of the Capitoline Temple, the *area Capitolina* or even only the temple itself.[84] However, the term can also refer to the entire hill, encompassing the Arx and the Asylum between the two peaks.[85] The size of the space differs substantially depending on which meaning is intended. Livy lends no clarity, as he uses the term in both senses. However, the senate traditionally met in the Capitoline Temple to discuss matters threatening Rome, and at the beginning of the consular year when military requirements were discussed.[86] The temple was also where consuls began and ended their campaigns under the sight of Jupiter Capitolinus.[87] It follows that the levy took place in the sacred space around the temple (*area Capitolina*), in order to invoke Jupiter's blessing on the upcoming campaign.

However, the change in meaning of *Capitolium* may also reveal the area of the levy. As the population increased over the centuries, it is possible that the gathering spread onto the relatively flat Asylum and even up the steep slopes of the Arx.[88] Thus the division of men into legions described by Polybius (further discussed later) would have still occurred in the *Capitolium* proper, but tribes would have waited outside until called.[89] If so, this suggests that while the *dilectus* continued to take place *in Capitolio*, the meaning of *in Capitolio* grew to encompass the necessary space. It is therefore worth including the area of both the *Capitolium* and the Asylum when considering space available for a levy on the Capitol.

---

83 MacMullen (1980) 455.
84 Platner & Ashby (1929) 95–8; e.g., Cicero, *Scaur.* 47; Varro, *Ling.* 5.149; Virgil, *Aen.* 9.448; Horace, *Carm.* 3.30; Pliny, *HN* 19.23.
85 Platner & Ashby (1929) 95–8; e.g., Cicero, *Font.* 30; Livy 3.18–19; Valerius Maximus 1.1.11; Suetonius, *Tib.* 3; Gellius, *NA* 17.21.24.
86 Weigel (1986) 333, 337.
87 Livy 45.39.
88 Cf. Coarelli (2007) 31.
89 Polybius 6.20.

A second, more problematic, difficulty is the buildings present in these spaces during the Middle Republic. Excavation of the *area Capitolina* has revealed several temples alongside the Capitoline Temple, which can thus be accounted for. However, there were numerous statues throughout the area, the number and situation of which are unknown. The several instances when statues had to be removed suggest a significant and hindering presence.[90] Alternatively, the need to remove statues may reveal that the *area* was considered overfull, but not to what degree. Rather than indicating congestion, it may simply refer to increasing difficulty in holding the levy and the tribal assembly. The Asylum is even more problematic. Other than the Temple of Veiovis built in 192,[91] a lack of evidence means that nothing is known about its features during the period under discussion. Therefore, the estimates presented in the following reflect the overall area only.

With the area under consideration established, the total capacity can be estimated. The Piazza del Campidoglio has an area of approximately 3339 m^2.[92] Taylor, Mouritsen and MacMullen allow four people per square metre when estimating crowd density. Following this, the total capacity of the piazza is 13,356.[93] Coarelli's map of the Capitol has been used to trace this space onto the area of the Asylum.[94] This results in an area of approximately 13,356 m^2 and a total capacity of 53,424 persons.

Within the *area Capitolina* the question is slightly more complex. The selection method described by Polybius, of bringing four men forward at a time of comparable age and fitness, suggests that space was required to make this selection.[95] If, as seems likely, the military tribunes and consul stood on the podium of the Capitoline Temple, the elevation would have made this process easier. Nonetheless, it is best to err on the side of caution here, assuming at least 1 m^2 per individual. The other limitation is that the men needed to be visible from the podium. Thus, only the area before the podium will be considered, and the areas hidden by the Temples of Fides and Ops Opifera excluded. On this basis, two piazzas fit in the space, giving a conservative estimate of 6678 m^2. This figure could be increased by about 25% and remain within sight of the podium, but as the *area Capitolina*

90 Livy 40.45.3; Piso, *FRHist* F40; cf. Cassius Hemina, *FRHist* F43.
91 Livy 31.21.12, 35.41.8 with Briscoe (1973) 113; Albertoni (1999) 99; Coarelli (2007) 39–40.
92 The point here is to generate an order of magnitude rather than an exact figure. The possibility of slight error will not have a substantial effect on the conclusions drawn.
93 Mouritsen (2001) 19; Taylor (1966) 113; MacMullen (1980) 454. This crowd density is within modern safe limits. 5 per m^2 is the maximum safe crowd density. Still (2014) 27–64. For animations see www.gkstill.com/Support/crowd-density/CrowdDensity-2.html.
94 Coarelli (2007) 28. The scale provided by Coarelli has not been used as it does not match up with the measurements obtained by empirical evidence.
95 Polybius 6.20.

contained an unknown number of statues, it is perhaps best to remain with the smaller figure. Thus 6678 spaced men, or 26,712 closely packed, could fit into the *area Capitolina*.

This conclusion can be tested against ancient evidence. Livy reports that in 167 the soldiers of L. Aemilius Paulus filled the Capitol with such a crowd that no one else was able to approach to vote for the consul's triumph.[96] Again, the use of *Capitolium* is ambiguous, but as tribal voting took place within the *area Capitolina*, it is reasonable to assume that the soldiers had filled this space.[97] Paulus commanded two legions, each of 6000 foot and 300 horse.[98] Additional men from the 168 levy were placed on garrison duty in Macedon.[99] Their number is unclear, but it is possible these men had also returned to Rome. Further, MacMullen notes that the army returning from Illyria was also in Rome.[100] These two legions serving under praetor L. Anicius Gallus totalled 11,000 men.[101] It is unclear if these men were also on the Capitol during the vote, but it is not unreasonable that they could have been. Anicius had conducted a related campaign against Macedon's Illyrian allies. The treatment of Macedon and Illyria in an almost identical manner by the senate, and almost as one by Livy, indicates the close links between the campaigns.[102] Moreover, Paulus had claimed spoils from those Illyrian towns which had supported Macedon. He had to send a letter to Anicius to prevent disturbance, presumably among the men, emphasising that there may have been friction between Paulus and the Illyrian legions.[103] Thus it is possible that 23,600 soldiers were in the *area Capitolina*. This is close to the 26,712 calculated earlier, suggesting that it is a reasonable estimate of the space available for gatherings on the Capitol.

Numbers at the dilectus

The second step in establishing whether the *dilectus* could have taken place on the Capitol is estimating the number of men required to present themselves for the levy. Previous attempts to calculate the number of *assidui* (those over the property qualification and thus liable for ordinary military service) in different years have been made, by Brunt and Rosenstein in particular.[104] However, as a slightly different question is at issue here, it is

---

96 Livy 45.36.6 – *postero die milites tanta frequentia Capitolium compleuerunt, ut aditus nulli praeterea ad suffragium ferendum esset*; Plutarch, *Aem.* 31.1–2.
97 Livy 25.3.8–14, 33.25.7, 34.1.4, 34.53.2, 43.16.9, 45.36.1.
98 Livy 44.21.8.
99 Livy 44.21.5–8.
100 MacMullen (1980) 456 n. 16.
101 Livy 44.21.10.
102 Livy 45.18.
103 Livy 45. 33.8–34.1.
104 Brunt (1971) 64–6; Rosenstein (2002) 184–6.

worth examining again. Brunt based his calculations on two assumptions: the size of the population and the ratio of *iuniores* (aged between 17 and 46) to *seniores* (aged between 47 and 60). This discussion aims to create a more accurate population model, basing any necessary assumptions on the Coale–Demeny[2] model life tables. The model created, and subsequent conclusions drawn from it, will thus be demographically plausible even though the exact nature of the Mid-Republican population is unknown.[105]

It is necessary to emphasise that quantifications made about ancient demography can never be precise. There are not enough data to make them so. Rather, they serve to provide an indication of scale. The Coale–Demeny[2] tables are models, based on probabilities for hypothetical stable populations. (In a stable population, the distribution of individuals across the different age categories remains the same even if the overall population changes.[106]) Any conclusions drawn from calculations based on these tables are only estimates, even where they appear very precise. Nonetheless, the sense of scale provided by these estimates opens a window onto the functioning of ancient societies.[107] Used with care, they aid in understanding the *dilectus*.

Two Roman census figures will be used in this discussion: the 234 figure of 270,212 and the 164 figure of 337,022.[108] Although they are found only in the *Periochae* of Livy, these figures are largely uncontested in modern scholarship. The population size they represent is debated, but the figures themselves are considered accurately transmitted. The 234 figure was chosen because it reflects the population prior to the Hannibalic War. Using a return from during the war, such as from 209,[109] would create problems because the death rate among the *iuniores assidui* was significantly higher than usual.[110] In effect, the population was no longer 'stable' and thus the census figures cannot be applied to the model life tables in the same way. This earlier figure also avoids the problem of Capua losing its Roman citizenship during the war; the number of citizens this entailed can only be estimated.[111] (The Polybian manpower figures of 225 are not used, as they are not census figures; see II:iii.) The 164 census figure is the highest

---

105 The author would like to thank April Pudsey for her comments on the following discussion.

106 Newell (1988) 120.

107 Newell (1988) 118; Parkin (1992) 68, 80–3; Saller (1994) 47; Salleres (2002) 1–5, 160–7; Hansen (2006) 1; Akrigg (2011) 47–57; Holleran & Pudsey (2011) 12–13; Hin (2013) 105–18.

108 [Livy], *Periochae* 20.15, 46.7.

109 Livy 27.36.6–7.

110 The reliability of casualty figures in the ancient literature will be discussed in III:ii. However, whatever view is taken on this subject, it is clear that the number killed during the Hannibalic War was far in excess of previous wars.

111 Cf. Brunt (1971) 64–6.

recorded during the Middle Republic; it provides a highest-possible-numbers scenario. It also reflects the population size shortly before statues were cleared from the *area Capitolina*.[112] If it was possible to hold the levy on the Capitol at this time, this figure may help demonstrate the space available.

Exactly what these census figures represent is a more difficult question. The census was begun in order to create a record of all those capable of military service, that is, the *iuniores* and *seniores*.[113] Thus the most common census-figure interpretation is as a record of all male citizens over 17 years old.[114] However, taking into account the census' role in taxation, Hin suggests that the census in fact reflects all those *sui iuris*.[115] This would include not only the appropriate *iuniores*, *seniores* and *senes* (men over 60), but also widows and orphans. As will be discussed later (II:i–ii), the *paterfamilias* provided the census declaration for his whole family.[116] This included any sons and grandsons for a military register. Hin suggests that different lists were compiled for different purposes, a suggestion with which this author agrees (I:v).[117] Despite this, it is unlikely that the published census figure, under the formula *censa sunt civium capita* (the heads of the citizens were counted),[118] changed from its traditionally ascribed role of enumerating male adult citizens.[119] Thus this interpretation of the census figures (all male citizens aged 17 and over, both above and below the property qualification for ordinary service) will be used here.

Another problem is under-registration in the census. Estimates for this range from 10% in the Middle Republic to 25% by the late first century.[120] What this means for estimating the citizen population has been hotly debated.[121] However, this is not such a problem as it initially appears. The purpose here is to understand the *dilectus*, not Italian demography. It is unlikely that those who avoided the census would then present themselves for the levy. Indeed, their absence would not have been noticed if the levy used lists derived from the census (see later). Moreover, under-registration figures are speculative, often used to support particular demographic

---

112 In 158, Piso, *FRHist* F40.
113 Livy 1.44.
114 Frank (1924) 329; Lo Cascio (1999) 163–4; De Ligt (2007) 121–4.
115 Hin (2008) 202–3.
116 See e.g., Toynbee (1965a) 445; Brunt (1971) 15; Nicolet (1980) 68; Lintott (1999) 117–8; Briscoe (2012) 434; cf. II:i–ii.
117 Hin (2008) 214.
118 Livy 10.47.2, 27.36.6–7, 35.9.2, 38.36.10, 42.1.2; [Livy], *Periochae* 10.10, 11.1, 13.7, 14.5, 16.5, 18.6, 19.7, 20.15, 27.22, 38.7, 41.8, 42.9, 45.9, 46.7, 47.7, 48.2.
119 Census-taking will be discussed in greater detail in II.
120 Brunt (1971) 35; Lo Cascio (2001) 123.
121 E.g., Frank (1924) 329; Brunt (1971) passim; Lo Cascio (1999) 163–4; De Ligt (2007) 121–4; Hin (2008) 202–18.

reconstructions. It is safer to use the figures as transmitted by the *Periochae* rather than attempt to alter them.

The *proletarii* are the final problem. *Proletarii* were men who fell beneath the lowest property qualification for ordinary military service but were liable in emergencies. Attempting to explain recruitment difficulties during the Hannibalic War, Brunt argues that *proletarii* were approximately 50% of the population in the late third century.[122] However, Rosenstein convincingly concludes that 10% is a closer approximation, arguing that the requirement for a man to remain on the family farm made up the shortfall felt by the Roman levy.[123] Therefore, in order to account for the *proletarii* here, a conservative 10% will be taken from the total of each age category in the population breakdown.[124]

In order to understand the population, the Coale–Demeny[2] model life tables will be used in conjunction with Saller's model population simulations. Following Parkin and Saller's considerations of likely life expectancy at birth (20–30 years), the West tables will be used.[125] The West tables are recommended for populations with statistics too poor to attribute them to the North, East or South models. Level 3 West Male and Level 6 West Male will be used.[126] These use a life expectancy at birth of 22.852 years and 30.073, respectively. As these ages fall at either end of the estimated 20–30 years' life expectancy, they provide a 'best'- and 'worst'-case scenario for numbers at the levy. Although the tables do not certainly replicate the Roman population, they are 'unlikely to be grossly misleading'.[127]

Using the model life tables, a breakdown of the men in each category was calculated in order to establish the number of *assidui* for each of the two models and census years. The process is outlined in Appendix I. These figures were then applied to the model population simulations run by Saller. Saller took two different average ages of marriage, which he refers to as 'ordinary' (women aged 20 and men aged 30) and 'senatorial' (women aged 15 and men aged 25).[128] Here the tables' 'proportion having living kin' and 'mean age of living kin' have been amalgamated to examine three situations: Level 3 West Male 'ordinary' marriage, Level 3 West Male 'senatorial' marriage and Level 6 West Male 'senatorial' marriage for both 234 and 164. Saller did not provide a simulation for

122 Brunt (1971) 66; Livy 22.11.2, 22.57.9, 23.14.2–4, 25.5.6–9; Appian, *Han.* 5.27.
123 Rosenstein (2002) passim. See also Rathbone (1993) 145–8.
124 As Northwood (2008) 269 argues, there is no reason to believe that the census regularly missed large numbers of *proletarii*. As they were not liable for regular military service or *tributum*, they had no incentive not to register.
125 Parkin (1992) 80; Saller (1994) 25.
126 Coale & Demeny (1983) 107, 110.
127 Saller (1994) 23.
128 Saller (1994) 45.

Level 6 West Male 'ordinary' marriage, so this cannot be examined. This range of examples provides a series of likely termini, from which conclusions about the numbers required to attend the levy can be drawn.[129]

Before continuing, the conditions imposed on the calculations must be discussed. The tables here were created on the assumption that the levy allowed one man aged 17–60 to remain on the farm. At 60 a male Roman citizen passed from the *seniores* into the *senes*, relinquishing any obligation to fight in Rome's defence.[130] It is reasonable to assume that the age at which Rome considered its men no longer physically able applied to farm labour as much as military service. Farming, particularly ploughing, was a labour-intensive occupation requiring considerable strength and stamina.[131] Additionally, Rosenstein has demonstrated that it was extremely rare for anyone over 35 to be enlisted.[132] He considers the 214 recruitment problems the result of labour requirements and a marriage pattern where men married at approximately age 30. This effectively resulted in a blanket exemption for married men, as well as sons of widows and possibly oldest sons of elderly parents. As a result, the number of those liable and able to serve under normal conditions was roughly 50% of the *assidui iuniores*.[133]

These exemptions may have been made during the census rather than by the consul during the levy. Exemptions could and did occur during the *dilectus*: the evidence reveals consuls issuing exemptions and plebeian tribunes involved in disputes.[134] However, there is also evidence of exemption at an earlier stage. Scholars have commented that Polybius' levy description fails to mention an exemption process.[135] Absence of evidence is not evidence of absence, but it may be that the majority of exemptions were not part of the *dilectus* and were thus not mentioned. More securely, Polybius calls those who came to Rome στρατευσίμοι, the fit-for-service. This implies that the unfit-for-service were exempted previously. Certainly it would have been impractical and extremely inefficient to issue an exemption every year

---

129 The Coale-Demeny² tables and Saller's simulations use age brackets which do not correspond exactly to the age of *assidui iuniores*, 17–46. Therefore the following calculations were done using both the 15–19 bracket from the tables, and a 17–19 bracket (given in parentheses) following the rationale detailed in Appendix I n.2. The proportion of the population within the 15–19 bracket is large enough to have a significant effect on the conclusions here. Using both brackets allows a more effective application of the model. The 45–49 bracket has not been included because the number of men recruited from this bracket was negligible; the number is more than compensated for by including the entire 15–19 bracket.

130 Varro ap. Nonius 523.24.

131 White (1970) 173–8, 194–5; Rosenstein (2004) 73.

132 Rosenstein (2004) 85–89; Livy 22.11.9.

133 Rosenstein (2004) 89; Livy 24.18.

134 E.g., [Livy], *Periochae* 48, Appian, *Hisp* 9.49.

135 Walbank (1957) 699; Brunt (1971) 628; Nicolet (1980) 98–9; Sage (2008) 122.

Table I.1 Liable and available *assidui iuniores*, 15–44 (17–44)

Model	234	164
Level 3 West Male 'ordinary' marriage	81,670 (56,969)	101,861 (71,053)
Level 3 West Male 'senatorial' marriage	63,052 (50,114)	78,643 (62,506)
Level 6 West Male 'senatorial' marriage	60,537 (47,997)	84,465 (66,969)

for the unsuitable, the maimed and those who had completed their service. A note taken during the census would solve this problem. It is not a great jump, given the preference of enlistment demonstrated by Rosenstein, to imagine that those married or the sole male labourer could receive an exemption in a similar way.[136]

Finally, the wording of the oath established by the censors in 169 again suggests a blanket exemption for veterans. This oath required all *iuniores* to swear that they would attend the *dilectus* if they had never served.[137] It makes no mention of service length. This may be a reflection of the increasing length of second-century campaigns. However, the oath was introduced within a context of recruitment difficulties and indiscriminately applied to all *iuniores*. This suggests that a large portion of *assidui* manpower was untapped, perhaps due to just these types of marriage and labour exemptions. Consequently, all married men, and men without an unmarried brother or a father under 60, are considered exempt for this examination.

The demographic evidence was analysed following the parameters established earlier. The results are summarised in Table I.1.[138]

The results show surprisingly little variation between the different marriage and mortality models. As they represent both ends of a plausible spectrum, this supports the conclusions drawn in the following. It should be noted that these figures give a model of the liable and able population using only exemptions based on marriage and labour requirements. In reality, exemptions for served terms, illness and maiming would also have lowered the number of available *iuniores assidui*. There is no evidence to gain any sense of scale for these exemptions; nevertheless, collectively they must have been relatively significant, especially with a service term of six years (see I:1). Further, the estimate of 10% *proletarii* is conservative; a larger estimate would further

---

136 The role of censor included assigning status. The census was originally a military review, and retained this aspect particularly with regard to the equestrians. As assessors of military fitness and category, it would be surprising if the censors did not have the power to grant exemptions.

137 Livy 43.14.5–6. See n.58.

138 The tables for the different age categories across the Level 3 West Male 'ordinary' marriage, Level 3 West 'senatorial' marriage and Level 6 West Male 'senatorial' marriage for the 234 census of 270,212 and the 164 census of 337,022 are fully laid out in Appendix I, Tables A1.3a–A1.5b. An example of the methodology is also provided.

lower the totals. Finally, the 15–44 models include a large number below the age qualification. Thus it is plausible that the true number of men liable and able to attend the *dilectus* was lower than that modelled here.

Discussion

These figures can now be compared with the Capitoline's capacity in order to assess whether it was possible to hold the levy *in Capitolio*. The Asylum's capacity up to the boundary of the *area Capitolina* was 53,424, and that within the *area Capitolina* at least 6678 widely spaced or 26,712 closely packed. All the 17–44 figures for 234 could have been contained *in Capitolio*. The 164 figures suggest that the *area Capitolina* was more densely packed than the lower estimate allows. However, it was previously demonstrated that the Macedonian and Illyrian legions could fit in the *area Capitolina* to vote in the tribal assembly. This suggests that although it may have been more difficult than usual, the levy could still take place on the Capitol. The crowd may also have started to overspill the flatter area of the Asylum onto the slopes. Moreover, it is probable that population increase was part of the reason for moving the *dilectus* to the *campus Martius* by the first century.[139]

The 15–44 model results are more problematic, but not insurmountably so. As already mentioned, the totals are overestimates, because they include 15- and 16-year-olds. Further, it is possible to increase the crowd density on the Capitol without reaching dangerous levels.[140] With a density of five men per square metre, the modern maximum safe level, the Asylum holds 66,780 and a packed *area Capitolina* holds 33,390. As with the 17–44 model, the number required to attend the *dilectus* in 164 would not easily fit within the defined area. However, the same caveats also apply. The physical evidence combined with the range of probable population models demonstrates that it was physically possible to hold the levy *in Capitolio* in the Middle Republic.

In light of this, the literary evidence can be reevaluated. Polybius and Livy are not alone in referring to the levy on the Capitol; Varro, Plutarch and Valerius Maximus do likewise.[141] Previously these references have been dismissed by scholars as a remnant of an older tradition referring to a sacrifice, the local levy for Rome itself or the division of the men into legions

---

139 Nicolet (1980) 100 using Dionysius, *RA* 8.87.3–5 suggests that this move was at least partly in order to remove the influence of the plebeian tribunes on the levy, as the *campus Martius* was outside their jurisdiction. This is plausible, but the increasing size of the liable and able *assidui* may have been a more pressing issue.

140 See n.93.

141 Polybius 6.19.6; Livy 25.3.8–14; Varro, *ap.* Non. 28L; Plutarch, *Marc.* 23.1 with Livy 26.31.11; [Livy], *Periochae* 14.3 and Valerius Maximus 6.3.4 with Varro, *Ling.* 6.86. Also referred to in Rome more generally: Livy 22.2.1, 25.3.4, 39.29.10, 41.5.4.

only, because Polybius' account missed a step.[142] These suggestions are not entirely implausible, but they require a nonliteral interpretation of the evidence. For example, Plutarch describes Marcellus performing a sacrifice on the Capitol in 210, but not the levy.[143] However, Livy describes Marcellus climbing the Capitol to perform the levy: *in Capitolium ad dilectum discessit*. It is difficult to otherwise interpret Livy's unambiguous statement.[144]

Livy's statement does not rule out a local levy, but it is unclear how Marcellus could have overseen a local levy from the central location. Additionally, citizens were able to appeal to the consul for an exemption during the levy process (even though, as already shown, the majority of exemptions occurred during the census).[145] If the *dilectus* was undertaken locally, the only opportunity for appeal was at the point of division, after the selection stage.[146] This would have resulted in understrength legions before even leaving Rome. Such a reconstruction leaves unclear how (or whether) replacement legionaries were recruited. Another logistical problem is coordinating a local levy. Mid-Republican Rome is generally considered a glorified *polis*.[147] How was it ensured that the correct number of men were chosen from the allotted tribes across different regions? By the first century this problem appears to have been surmounted,[148] but there is nothing in Polybius' description or the scattered references across other ancient writers to suggest such a mechanism during the Middle Republic.

Brunt suggests that Rome's response to Hannibal's attack on Rome in 211 is evidence for local levies, arguing that Polybius' account of 211 undermines his levy narrative.[149] Rome encountered a stroke of luck when Hannibal arrived outside the gates. The consuls had instructed the first army enrolled to present themselves at Rome, in arms, on that day. They were also engaged in sorting and examining a second army, τὰς καταγραφὰς ἐποιοῦντο καὶ δοκιμασίας. However, this does not suggest a localised levy. Rather, the first army were completing the final mobilisation step as set out by Polybius: reassembling for a second time fully armed and ready for campaign.[150] This is the step most frequently attested by Livy.[151]

---

142 Walbank (1957) 698–9; Brunt (1971) 628; Sage (2008) 121.
143 Plutarch, *Marc.* 23.1. The sacrifice may be connected to auspices.
144 Livy 26.31.11
145 E.g., [Livy], *Periochae* 48; Appian, *Hisp.* 9.49.
146 Polybius 6.20.
147 Scheidel (2004) 6.
148 The Spanish, Caesarian *Lex Ursonensis* in particular demonstrates that a local levy took place at least in the provinces by the mid-first century; *CIL* II.5439.
149 Brunt (1971) 627–8; Polybius 9.6.6.
150 Polybius 6.26.2.
151 E.g., Livy 22.11–12, 23.31.5, 23.32.2, 34.8.4–5. Nicolet (1980) 102 is wrong to suggest that this reassembly is at a local enrolment centre, see below.

The second army could be undergoing the selection stage of the levy, but, as Brunt himself points out, if Polybius were describing all the *iuniores* as present in Rome he would have expressed himself differently.[152] Polybius explicitly states that Rome is lucky because of the presence of two nearly fully mobilised armies. It appears that the second army was reassembling for the first time, the point at which the men were divided into battle lines.[153] As this passage helps to clarify, it was at this stage that the examination of age and wealth (often considered missing from the account of initial selection) took place. The division appears as one of relative age and wealth, rather than a carefully calculated decision earlier in the process. The Capitol is not mentioned because this process likely took place outside the *pomerium* on the *campus Martius*. Thus, the response to Hannibal's advance on Rome is not evidence of a local levy. Rather, it illustrates the later stages of legion organisation, which (unlike the levy itself) could be undertaken legion by legion.

Events caused by the censors in 169 have also been used by Brunt to support the local levy.[154] In addition to introducing a new oath for *iuniores* who had not served (see earlier), the censors ordered that all veterans discharged from the Macedonian campaigns since 172 present themselves to the censors for a dismissal review. This resulted in a huge throng of men in Rome, much greater than usual.[155] Brunt argues that the crowd's size was unusual because men had been summoned to Rome who were usually levied locally. This requires *hoc edicto* to refer to only the decree concerning the Macedonian veterans. However, the narrative dealing with the censors' action is constructed as a single block. The concluding *hoc edicto* refers to the actions in the singular sense of 'the work of the censors'. The unusual crowd at Rome was a result of both aspects of the censors' work. The crowd's size was in comparison to the dwindling turnout at the *dilectus*, fuelled by the generous exemption grants of the consuls. Through interpreting *hoc edicto* in this way, the passage becomes further support for the levy in Rome.

A final piece of evidence to be discussed is that of six commissioners being sent out from Rome to aid in the 212 levy.[156] This is a rare example of the levy occurring outside Rome.[157] However, the levy's circumstances were noteworthy precisely because they were unusual. The commissioners appear to have enlisted any fit, free man on the spot where he was found, *in pagis forisque et conciliabulis*. (The text's implication is that only the first

---

152 Brunt (1971) 628.
153 Polybius 6.21.6–10.
154 Brunt (1971) 633–4.
155 Livy 43.14.7–10 – *hoc edicto litterisque censorum per fora et conciliabula dimissis tanta multitudo iuniorum Romam convenit, ut gravis urbi tanta insolita esset.*
156 Livy 25.5.6–9.
157 Cf. Livy 7.25.7–9, 24.20.13, 32.26.12.

stage of the levy, selecting men to serve, was undertaken by the commissioners. There is no reason to believe that the second stage did not occur as normal.) This levy has much more the character of a *tumultus*, an emergency levy in which exemptions were ignored, although Livy does not use the term here.[158] It must be conceded that the locations listed, particularly *fora* and *conciliabula*, are where a local levy might be imagined to occur. However, that two sets of triumvirs were specially created to perform this duty suggests that a levy outside Rome was highly unusual, the result of the Hannibalic War's extreme need, and not normal practice.

This is emphasised by how the triumvirs were assigned. One group was given the area within 50 miles of Rome, and the other that beyond.[159] On foot, 50 miles is roughly a two-day journey; some citizens lived more than double this distance from Rome. It is at this kind of distance that a local levy might be expected to be ordinary practice. However, that triumvirs were sent out this far indicates that no such local organisation was in place. Indeed, any local levy would likely have also operated on a tribal basis. By the Middle Republic, members of the same tribe lived in different areas.[160] This would require a great deal of travel for those living furthest from Rome, as it is unlikely that the consuls would go any great distance from the city. From this perspective, travelling a few extra miles to reach Rome itself would not have been a much greater journey.

Overall, both the topographical and literary evidence demonstrates not only that Polybius' description of the *dilectus* was accurate in placing the levy *in Capitolio* but also that his description has not missed out a previous stage. As holding the levy on the Capitol was physically possible, there is no need to consider the scattered literary references to the practice anachronistic or mistaken. For this discussion, it is significant that holding the levy on the Capitol required written records. Granting or recording exemptions with the census declaration suggests that some form of personal service record was held by Rome for each citizen.[161] Thus the lists of those liable for military service created for the *dilectus* were most likely closer in length to the totals generated from the model populations already

---

158 Livy 7.9.6 (*tumultus Gallici causa, omnes iuniores sacramento adegit*), 7.28.3 (*tumultus, dilectus sine vacationibus habitus esset*), 32.26.12 (*sacramento rogatos arma capere, tumultario dilectu*) 34.56.11 (*tumultum esse decrevit*), 35.23.8 (*tumultariorum*), 37.57.5 (*tumltario exercitu collecto*) 40.26.7 (*tumultarios scriberet*), 41.5.4 (*in tumultu*), 43.11.10 (*tumultario dilectu conscriptos*); see II:iii. Golden (2013) 42–86 argues that *tumultus* often occurred without the term being used by ancient authors but can be identified by actions taken.

159 Livy 25.5.6–9 – *alteros qui citra, alteros qui ultra quinquagensimum lapidem.*

160 Sergia, Clustumina, Claudia, Pollia and Sabatina became divided tribes during the first half of the third century, and possibly Papina, Voltinia and Oufentina as well, Taylor (1960) 68–100.

161 The question of where such records would have been gathered and stored will be addressed in V.

examined (Table I.1) than to the census figures. Levy centralisation was easier from an administrative standpoint because recordkeeping was also centralised (see V:iii). In order to examine the operation of these records in more detail, it is necessary to investigate the next element of Polybius' narrative.

v Selecting *milites*

The use of written administration becomes clearer with the selection of tribes and individuals to fill the legions. Polybius describes the selection of tribes by lot, followed by the distribution of men into the legions:

> ... κληροῦσι τὰς φυλὰς κατὰ μίαν καὶ προσκαλοῦνται τὴν ἀεὶ λαχοῦσαν. ἐκ δὲ ταύτης ἐκλέγουσι τῶν νεανίσκων τέτταρας ἐπιεικῶς τοὺς παραπλησίους ταῖς ἡλικίαις καὶ ταῖς ἕξεσι.
>
> (Polybius 6.20.2–3)

> ... they order tribes by lot one by one and summon every one as drawn by lot. From this they pick out four of the young men suitably matched in age and bearing.

The passage goes on to describe each of the four groups of military tribunes choosing in rotating turn one of the men brought up in fours.[162] This created legions of heterogeneous composition, making their strength relatively equal as well as reproducing the structure of Roman society in miniature.[163]

The selection of men by tribe is attested elsewhere.[164] In 275, for example, no one appeared on the Capitol on the appointed day. The consul Curius Dentatus resorted to reading names from the list of the allotted tribe.[165] Importantly, his ability to do this implies that he had a list of the liable citizens from that tribe.[166] As the census was conducted tribe by tribe, creating tribal lists of the liable is logical. Taylor suggests that the tribal lot

---

162 The levy in the archaic period is beyond the scope here. However, a fifth-century tribal levy is visible in the (not unproblematic) evidence, suggesting that the use of writing in the levy dates to at least the early Republic. Cf. Dion. Hal., *Ant. Rom.* 4.16, 4.19; Livy 4.46.1; Ogilvie (1965) 604; Taylor (1957) 341; Thomsen (1980) 188; contra Gabba (1951) 251–2.

163 Helm (2020).

164 Appian (*Hisp.* 9.49) attests the first instance of selecting men (not tribes) by lot in 152. He does not provide more detail. Like the 275 incident, it is presented as a reaction to recruitment problems and may have been an isolated case. Regardless, it again demonstrates the importance of documentation to the levy.

165 [Livy], *Periochae* 14.3; Valerius Maximus 6.3.4.

166 Cf. Taylor (1957) 343; Brunt (1971) 631.

was conducted with inscribed wooden balls shaken or swirled from a water carrier.[167] Polybius gives no indication of how tribes were allotted, but the weight of evidence provided by Taylor and lack of contradiction means there is no reason not to accept her conclusion.

Livy repeatedly mentions the giving and taking of names at the *dilectus*, but this is notably missing from Polybius' account.[168] When would such name calling most likely have occurred? Southern states that the names of the enrolled were recorded at the first reassembly.[169] However, this is implausible. It is unclear, in her reconstruction, how officials would have known who should have attended the reassembly. The situation necessitates a list of those enrolled, although possibly this was just marked against the tribal list. The point at which each citizen was personally selected by the military tribunes is the most obvious place in Polybius' account for this to happen. Names may well have been called for the division, but at a secondary stage following the original drawing up of a legion list or similar on the Capitol. Polybius' failure to mention the taking of names is probably because he considered it too obvious to need to include; as mentioned previously (I:ii), a list is implicit in his language.

Further, at the first reassembly the men were examined for relative age and class to be divided into battle lines (see I:iv). The presence of the tribal list which contained this information would be prudent. Polybius implies that not only age but also experience was instrumental in this division, indicating that the tribal lists contained this information.[170] No location for this assembly is explicitly mentioned; Polybius instead notes that a place was specified by the military tribunes.[171] As seen previously, this place could be Rome but was not necessarily so. To have the information on hand for the division into lines, a legion list of the enrolled which could be taken to the assembly place was required. As men from the same tribes were split across different legions, lists separate from the tribal lists used at the levy were required. This was particularly true if the musters occurred in different places at different times. Thus it appears that during the levy on the Capitol, legion lists were generated from the tribal lists which included details of service length, age and property qualification.

167 Taylor (1966) 71–2 using Plautus, *Cas.* 295–428 with *Tabula Hebana* 23, Horace, *Carm.* 2.3.25–27, Lucan 5.394; Oliver & Palmer (1954) 229, 239–41.
168 E.g., Livy 10.4.1–3, 10.25.2, 27.46.3, 37.4.3.
169 Southern (2007) 92.
170 Polybius 6.19–26 passim, especially 6.24.1 – Ἐξ ἕκαστον δὲ τῶν προειρημένων γενῶν πλὴν τῶν νεωτάτων ἐξέλεξαν ταξιάρχους ἀριστίνδην δέκα. This implies that many men in the *hastati*, *principes* and *triarii* had previous experience, and perhaps had been selected due to this. The importance of campaigns served in military and political office (see I:iii) also suggests that tribal lists contained information about past service.
171 Polybius 6.21.

vi Legion size

Following his description of the selection process, Polybius notes the size of an ordinary and emergency levy. This section highlights several points of interest in the discussion of both Polybius' reliability and the use of written administration in the *dilectus*.

ὅταν δ᾽ ἐκλέξωσι τὸ προκείμενον πλῆθος – τοῦτο δ᾽ ἔστιν ὁτὲ μὲν εἰς ἕκαστον στρατόπεδον πεζοὶ τετρακισχίλιοι καὶ διακόσιοι, ποτὲ δὲ πεντακισχίλιοι, ἐπειδὰν μείζων τις αὐτοῖς προφαίνηται κίνδυνος – μετὰ ταῦτα...

(Polybius 6.20.8)

When they had chosen the prescribed number – that is when in each legion there are four thousand and two hundred foot, and sometimes five thousand, whenever some greater danger should manifest to them – after this ...

The size of the legion described here by Polybius has been used as an argument against his reliability. Polybius is not consistent when giving legion sizes. In book 3 the 'standard' size of a legion is 4000 foot and 200 horse, and the figure of 4000 foot is again repeated later in the description of the levy.[172] However, the total number from the lines of *velites, hastati, principes* and *triarii* given is 4200, suggesting that mentions of 4000 are rounded from 4200. Sumner is right to consider 4200 foot and 300 horse the standard complement implied by Polybius.[173]

Ordinary legions of 4200 foot are considered to belong to the third century. It is generally argued that at some point during the Hannibalic War 5000–5200 became the standard legion size, increasing to 6000 at times of emergency.[174] Polybius is judged to have followed an older source (perhaps Fabius Pictor) without altering it to suit his own time.[175] Allowing that Polybius had some familiarity with the legions, Brunt attributes this error to a failure to account for casualties from battle and disease; 6000 men would appear closer to 5000, and 5000 closer to 4000, giving Polybius no reason to question or modify his source.[176]

However, there are problems with this interpretation. Firstly, as argued at I:ii, Polybius was not just familiar with the legions; he had been in close contact with them and was valued as a tactician. That he would have been

---

172 Polybius 3.107.10, 6.21.10.
173 Sumner (1970) 67.
174 Brunt (1971) 423, 467, 672–5; De Ligt (2007); Roth (1994) 347 is more nuanced; Toynbee (1965a) 506 is a notable exception; cf. Livy 44.21.8 for legions of 6000.
175 Brunt (1971) 675.
176 Brunt (1971) 675.

involved in producing strategy but unaware of operational size seems unlikely. Secondly, with the levy occurring in Rome it seems impossible that an interested Polybius would not have bothered to find out the size of the legions created on his doorstep.[177] Finally, the losses proposed by Brunt approach 20%. These are well above estimated averages.[178] When it is recalled that Polybius was summoned to Lilybaeum in 148 to newly formed legions who had yet to see service,[179] Brunt's interpretation becomes untenable. Polybius cannot be considered conscientious and his legion size simultaneously dismissed.

The senate formally decided the number of legions and men to be enrolled each year. Weigel has highlighted that these deliberations are not mentioned in the sources nearly as often as recruitment and the declaration of war.[180] Nonetheless, they are frequent enough to make it clear that legion size was prescribed by the senate.[181] The notion of a set 'standard' size is then perhaps a misnomer; rather, Polybius indicates there were 'standard sizes'. Roth suggests that 5000–5200 had become the customary legion size in the second century.[182] Certainly, Livy mentions legions of 5200 more often than 4200 in books 21–45.[183] However, he does not mention legion size for most years. When he does, it may be precisely because they were not 'standard'. The numbers recorded by Polybius are thus two of the standards (not prescriptions) applicable for the third and second centuries.

As noted already (I:ii), this is even clearer when the construction of Polybius' remarks is examined. As is emphasised in modern editions and translations, the section giving the number in a legion can be removed from the text without disrupting the narrative, grammar or syntax.[184] This suggests that even if Polybius did base his description on an older source (which the author hopes this discussion has demonstrated is not a necessary conclusion), there is no need to believe that he did so blindly. The whole narrative can function without it, bar a single reference to 4000.[185] This in itself shows that Polybius was creating a

177 See I:iv.
178 Rosenstein (2004) 109; see III:ii.
179 Polybius 36.11.1.
180 Weigel (1986) 334.
181 See e.g., n.181.
182 Roth (1994) 347.
183 E.g., Livy 21.17.2–3 (4000 foot, 300 horse), 23.34.13 (5000 horse, 400 cavalry), 26.28.7 (5000 horse, 300 foot), 29.24.11–14 (6200 foot, 300 horse), 32.8.2 (4000 foot, 300 horse), 32.28.10 (6000 foot, 300 horse), 35.41.5 (4000 foot, 300 horse), 39.38.10 (4000 foot, 300 horse), 40.1.5 (5200 foot, 300 horse), 40.18.5–6 (5200 foot, 300 horse), 40.36.7 (5200 foot, 400 horse), 41.9.2 (5200 foot, 300 horse), 42.31.2 (6000 foot, 300 horse), 43.12.3–6 (6000 foot, 250 horse; 5200 foot, 300 horse).
184 Paton, Walbank & Habicht (2011) 350–1; Waterfield (2010) 386.
185 Polybius 6.21.10.

coherent whole. It is clear that Polybius was envisaging a levy of four legions of 4200 men each, but his method could easily be applied to different-sized legions, as the note on the unchanging number of *triarii* demonstrates.[186] Indeed, as a writer aware of and interested in his surroundings, Polybius demonstrates that contemporaneous levies of both 4200 and more were familiar to him.

Finally, it is worth highlighting that any prescribed legion size required a record of the number enrolled. This does not necessitate more than a tally. Nevertheless, it was impossible to create a legion in the manner described by Polybius, or indeed of any preordained size, without taking at least a rudimentary record during the process. Although there is no explicit supporting evidence, it is not unreasonable to imagine that it was from this basis that *dilectus* administration developed to become the important fixture it was by 275. It appears that records became so fundamental to the process that they were not consciously considered by Polybius in his narrative. This demonstrates that the written element had developed into an integral part of raising Roman legions.

vii Paperwork and the *dilectus*

Overall, Polybius' military digression in book 6 should be considered a reflection of the contemporary situation in Rome. Issues such as the Capitol's capacity compared to the population liable for military service serve to demonstrate that Polybius' narrative can be accepted. As this is the major element highlighted as implausible by modern scholars, the importance of this demonstration should not be ignored. Rather than an at best anachronistic and at worst entirely implausible account written by an uninformed author, Polybius provides a detailed and informative narrative of second-century military structures. The notes on changes in form indicate that the account can also be used in the discussion of the third century. When considered in line with the mentions of the *dilectus* by Livy, it is clear that Polybius' narrative should be accepted in its entirety. Accepting Polybius' credibility opens an interesting window onto both the existence and functioning of a military bureaucracy.

The complex nature of the selection process of the *dilectus* as understood from the evidence of Polybius and Livy presupposes the existence of written administration. It may be questioned whether it is right or safe to assume that Republican Romans were operating a such a bureaucracy. However, given the apparent complexity of the written administration needed to deal with the processes in the evidence, it is difficult to understand what other method might have been utilised to keep this system in order. The census itself is an accepted example of bureaucracy during the Republic; it is

---

186 Polybius 6.21.10.

supported by substantial evidence and accepted by modern scholars as an important record despite the lack of any extant returns or rolls. It is not unreasonable, then, to posit other similar elements of bureaucracy inter-related with the census which aided the state in organising issues which were also the concern of the census.

The issue of exemptions in particular, coupled with an expected normal service term of only six years, highlights the level of bureaucracy. Exemptions took place not only during the *dilectus*, as is commonly accepted, but also during the census. Furthermore, the levy used tribal lists, either directly from the census or lists of the liable derived from it. This interaction of the two events and sets of magistrates indicates a level of bureaucratic interconnec-tion in the Middle Republic perhaps not always appreciated by modern scholars. Such record keeping indicates a state concerned with both under-standing its realistic manpower and not overburdening its citizens with military duty. This was motivated in part by pragmatic issues; both mutiny over long service and enrolment problems were not unknown. However, it is noteworthy that the system was organised in such a way that it was largely successful in preventing these problems in the first place.

Significantly, it is possible to trace a level of administration similar to that found in Polybius and Livy to the early third century. That it was in this form by at least 275 suggests that it developed from a much older origin. Furthermore, the terminology used by ancient authors in connection with the levy suggests that the involvement of written administration was so close to military organisation as to be indistinguishable, again indicating an ancient link. The nature of the army's composition may have changed, but the selection of soldiers became important as soon as a size was set. When the Roman army became organised in this way is uncertain, but it is rea-sonable to date the genesis of military bureaucracy to this period. Indeed, ancient authors themselves point out its very earliest form: the first census. Written administration, however simple its origins, was a fundamental and entwined part of the Republican levy.

The demonstration that written administration was not only present in the Mid-Republican levy but an integral part of it allows the scope of this investigation to widen. It is now possible to raise further questions. How far did this bureaucracy extend across an individual's military career? How detailed was the record keeping? Exactly what would have been recorded on the tribal lists? Did the Republican Romans create anything similar to what would today be recognised as a service record, whether a separate document or not? The emerging importance of the connection of the *di-lectus* and the census requires that the discussion of Roman military bu-reaucracy continue with an examination of the role of the census in recording military service.

2 The census and centralised military bureaucracy

The *dilectus* was not the only aspect of military organisation which required a record of liable and available men. More generally, Rome needed a reasonably accurate manpower record in order to mount campaigns and form legions as the senate instructed. This chapter aims to demonstrate that Rome continued to track her manpower after the enlistment stage. It examines the functions of this record keeping in Rome: the role of the census as well as more extra-ordinary methods of gaining accurate information about potential manpower in emergency situations. As already emphasised, the census was originally – and remained in the Middle Republic, to an important extent – a military review. As such, the census as an institution is of great importance for gaining a greater view of the nature of military bureaucracy in this period. It is widely recognised that Rome enumerated the men liable to fight for her through the census;[1] establishing how this was achieved is an important step in understanding Rome's military bureaucracy. It is argued here that this was accomplished by regular citizen declarations of military service at the census. Further, it is demonstrated that Rome was able to adapt to the situation at hand in order to keep manpower records as accurate as possible, whether during long service abroad or in more immediate emergency situations.[2]

i Census declarations

Censors could grant exemptions from service on the basis of, among other things, *emerita stipendia*, as could the consuls if a citizen appealed during

---

1 E.g., Bourne (1952) 130, 133; Toynbee (1965a) 445, 453, 462; Brunt (1971) 21–4.
2 The entire census system will not be examined, as it has no bearing on the specific facet of military records under discussion. For example, the property qualification is not discussed, because the exact wealth necessary to serve in the army had no effect on how the census itself functioned. Indeed, the author follows Rich (1983) 316 in believing that without further discoveries, the discrepancies in the evidence mean it is impossible to draw strong conclusions about the level of the property qualification at any point in the Republic. Contra, e.g., Marquardt (1891) 80; Brunt (1971) 402–6; Marchetti (1976) 154–6; Gabba (1976) 2–7.

the *dilectus* (see I:iii). Further, both Polybius and the *tabula Heracleensis* state that there was a minimum service requirement for holding a magistracy.[3] These facts suggest that the census record included a notation of the campaigns served by each man. This section adds to the conclusions of the preceding chapter, demonstrating that the work of the censors included noting the campaign years served and that the census declaration included a statement of, at the very least, how many campaign years had been served by each man over 17. This is a fundamental issue to establish if it is to follow that the census was the centre and beginning of tracking military manpower in Rome.

The proceedings of the 169 census are a good place to begin. Livy records that the census revealed a large number of absentees from the Macedonian legions, '*qui quam multi abessent ab signis census docuit*'.[4] Livy's language is emphatic. It is difficult to suggest that *docuit* (from *docere* meaning 'to teach, show or demonstrate')[5] with *census* as its subject does not refer to a record revealing the situation. That Livy is this emphatic concerning the role of the census and the clarity of the conclusions drawn from it indicates that the census was a detailed, carefully kept record.

However, the censors of 169 were already concerned with the Macedonian legions; Livy notes in particular the requirement that those with *potestas* over the men in the Macedonian legions appear to the censors in person.[6] It is possible that the census process, not the documents themselves, revealed the situation here. This cannot be entirely disproved, but several factors weigh against it. The number of men required to attend the censors themselves may have been slightly higher than the norm, but (as will be seen later) this number was not in itself indicative of absentee numbers. While the text does not give an exact figure, and could simply imply a sense of scale, 'how many' suggests that the censors themselves had a specific figure even if Livy (or his source) chose not to include it. Further, this number would have been relatively small compared to both Rome's entire population and eligible manpower, and thus not too difficult to calculate. Finally, as will be seen in III, it was in the interest of magistrates to enumerate as accurately as possible the men on campaign and in each legion. Thus, on balance, it can be concluded that Livy's account demonstrates that the census as a record revealed clearly to the censors what they had already suspected concerning the numbers absent from the Macedonian legions.

This has some interesting implications. Firstly, it suggests that the censors had a record to indicate who should have been serving with the legions, to

---

3 Polybius 6.19.1–4 (ten years of military service), *Tabula Heracleensis* ll.98–107 (three years of service in the cavalry or six years on foot), see I:iii.
4 Livy 43.15.7 – 'the census demonstrated how many were absent from the standards'.
5 *OLD* s.v. *doceo*.
6 Livy 43.14.7–10.

cross-check against those recorded in the census. This may well have been a copy of the legion list proposed in the previous chapter. Secondly, it demonstrates that military service was part of the census declaration. There is no need to follow Briscoe in concluding that the censors did not know who had been released from their term or had not fulfilled their obligations.[7] The problem the censors addressed was not unauthorised absenteeism but the premature discharge of men by politically motivated generals. Strictly speaking, all the men had been released from their terms and had thus fulfilled their obligations.[8] The censors wished to return those who had been, in their eyes, prematurely discharged. Those found to have been prematurely discharged were given 30 days to leave Italy, indicating that the censors allowed those who had legitimately gained *emerita stipendia* to return home.[9] Such a distinction would only have been possible had the censors possessed a service record to compare with the census declaration of the dismissed men (or their *patresfamilias*). Men who believed themselves legitimately dismissed had no reason to lie to the censors about their condition, but the censors could only have judged the situation if they had information beyond that given in 169's declarations.

What was the record which allowed the censors to act so? The clearest evidence of the composition of census declarations is provided by the *tabula Heracleensis*. The inscription is Caesarian in date, so care must be taken with its use.[10] The exact date of the measure itself is disputed, but modern scholars agree that it is no earlier than 88.[11] Further, the tablet represents a period when the census was taken locally, which, as will be seen (II:ii), was not the case in the Middle Republic. Nevertheless, it is still possible to extract relevant data. Frederiksen suggests that the tablet represents the raw material for a municipal charter, which was ordinarily tailored to fit the location.[12] If he is correct – and the reference to the Roman model in the text suggests he is – the record is extremely significant, as it indicates the process across Italy and possibly the entire empire. Moreover, Rome was slow to change processes. While the census location had changed, there is no reason to believe that the required form was substantially different, if at all, from that of the previous century. With these concerns in mind, the

---

7 Briscoe (2012) 438.
8 Livy 43.14.7 – *multos ex Macedonicis legionibus incertis commeatibus per ambitionem imperatorum ab exercitu abesse*.
9 Livy 43.14.8–10. This section of the description of the censors' actions seems to suggest that the review of discharges was a more general one than simply for the Macedonian absentees, but its actions would also have applied to them.
10 See II:iii for a more detailed discussion of the limitations of the *tabula*. These are taken as understood here.
11 E.g., Hardy (1914) 85; Reid (1915) 237; Toynbee (1965a) 457; Brunt (1971) 522; Nicolet (1980) 61; Lo Cascio (1990) 308, 312–3.
12 Frederiksen (1965) 197.

tabula Heracleensis can be used to shed light on the Mid-Republican census declaration.

According to the *tabula Heracleensis*, the designated magistrate received under oath

> *eorumque nomina praenomina patres aut patronos tribus cognomina et quos annos quisque eorum habe<bi>t et rationem pecuniae ex formula census*

> their names, their praenomina, their fathers or patrons, their tribe, their cognomina, and how many years old each of them shall be and an account of their property, according to the schedule of the census.[13]

Crawford et al. suggest that requiring the *nomen* first (rather than the *praenomen*) indicates that the census was performed in alphabetical order, providing as a comparison a Flavian list of *iuniores*.[14] This is a tempting theory, as it further supports the notion of an organised and complex bureaucracy. Deliberate organisation by name would suggest a record which could be easily accessed and checked for issues such as granting exemptions and military service in between *lustra*. During the Middle Republic this organisation would likely have been by tribe rather than by *municipium*, as in the tablet. However, it cannot be irrefutably stated that the lists were organised in this alphabetical manner. There is no Mid-Republican evidence either in support or against. It must be concluded that, however tempting the alphabetical hypothesis, the *tabula Heracleensis* demonstrates only the level of detail required for identification, not the broader method of composing the census.

The *formula census* does not include a mention of service length, in apparent contrast with the hypothesis proposed here. However, the *formula census* was sent from Rome,[15] allowing the presiding magistrate in Rome to add other criteria as necessary. This suggests that the *tabula*'s formula is the basic or usual census form, but that this form was not immutable. It also suggests that recording service campaigns was not usual, apparently demonstrating that questions regarding military service were not a regular feature of the first-century census.

On the other hand, first-century army recruitment, especially during the Civil Wars, became much more like that seen in the extraordinary

---

13 *Tabula Heracleensis* ll.146–7 (trans. Crawford *et al.* (1996) 377).
14 Crawford *et al.* (1996) 389 using *CIL* VI.200.
15 *Tabula Heracleensis* ll.147–8 – *quae Romae ab eo, qui tum censum populi acturus est, proposita erit.* 'Which will have been posted at Rome by him who is then about to conduct the census of the people'.

circumstances of 212, when commissioners enrolled men as they came across them.[16] The property qualification had also become much less important.[17] Consequently, it is less surprising to see the number of years served omitted from the Heraclean law. Indeed, perhaps a statement of age had become enough to signify eligibility for the army (although there is no direct evidence for this). The voluntary and more emergency nature of military service resulted in age being the only criterion for eligibility. Moreover, the lack of a full census between 69 and 29 indicates that census records may well not have been kept up to date, reducing and perhaps entirely eliminating the census' role in both recruitment and tracking manpower. Thus, while the tablet does illuminate facets of the census unavailable elsewhere, these conclusions cannot be fully accepted without also examining evidence relating directly to the Middle Republic.[18]

The Hannibalic War provides several illuminating examples regarding the taking and organisation of the census. The censors reduced *iuniores assidui* (in 214) and *equites* (in 209) who had not served during the war to the status of *aerarii*, 'taxpayers'.[19] Botsford argues that degradation to the *aerarii* was originally exactly as stated: removal from the tribes and placement in a special taxpayer category. After 304, when the lower classes, *humiles*, were limited to the four urban tribes,[20] the punishment appears to have come to mean movement into the aerarian class within one of the urban tribes. This class probably voted with the proletarian century, significantly reducing the political power of men moved to this category.[21] Lintott agrees with this view, highlighting that it was impossible for a censor to remove a citizen from the tribes.[22] The language used by Livy here supports this interpretation. *Tribuque omnes moti* allows for flexibility in the translation of *tribu*. However, the singular ablative suggests a

---

16 Livy 25.5.6–8, cf., e.g., Appian, *BCiv.* 3.40.
17 Marius' volunteer army of 107 should not be viewed as the watershed moment that it often is. The use of infantry volunteers is attested by Livy as early as 295 (Livy 10.25.1), and in 205 Scipio Africanus took an army of volunteers to Africa when he was refused permission to levy more troops (Livy 28.45.13–4). Marius' army may have been notable for the number of men below the property qualification, but it was in no way a huge break from the norm (cf. Gabba (1976) 11–2). Nonetheless, it is indicative of the increasing use of poorer volunteers over traditional recruitment.
18 It appears a problematic contradiction that the *tabula* mentions military service as a criterion for holding office but does not mention a method for its calculation. However, as this contradiction was not problematic for the tablet's author, and because it reflects a period outside this study's scope, there is no need to dwell on the issue further.
19 Livy 24.18.7–8 – *nomina omnium ex iuniorum tabulis excerpserunt qui quadrienno non militassent, quibus neque vacatio iusta militiae neque morbus causa fuisset. et ea supra duo milia nominum in aerarios relata tribuque omnes moti*; Livy 27.11.15.
20 Livy 9.46.14.
21 Botsford (1909) 62–5.
22 Lintott (1999) 118 using Livy 45.15.3–7.

translation of 'all were removed from their tribe', allowing for movement into a different tribe. If Livy had meant removal from the tribes entirely, the plural would be required. Being reduced to the *aerarii* was a change in census rating and tribe, and thus where the men were recorded, but not removal from the tribes altogether.

More important here is how this change was effected. Livy describes the censors picking out offenders from the table of *iuniores*, '*nomina omnium ex iuniorum tabulis excerpserunt*'. The *tabulae iuniorum* appear to be derived from the census and correspond with that proposed in the previous chapter as organised by tribe (I:v).[23] The use of *excerpo* here carries the double meaning of both selection and removal,[24] indicating that as well as identifying the individuals, the censors also removed them from this list. Further, that men who had not served since 218 without an exemption could be identified suggests that the *tabulae* (and thus the census list from which they were derived) included a service record. For the 214 example, it is possible that this information was simply taken from the census declarations of 214. However, given the crisis of 216 and subsequent recruitment problems, it is unlikely that men who had avoided enlisting volunteered that knowledge. Thus it appears that a longer-term approach to record keeping was required to keep track of Rome's manpower. This is confirmed by the censors' ability to do the same in 209.

The year 209 also provides another illustrative example of how the census functioned, on this occasion with regard to the *equites*. During the census, the survivors of Cannae with a public horse had it removed and were deemed to owe ten years of cavalry service.[25] At each census, holders of a public horse presented themselves to the censors and recounted their deeds.[26] This was not a review of all who reached the property qualification for cavalry service, but of an elite group honoured with a horse at public expense. *Equites* serving on their own horses is attested in the early fourth century, when volunteers served as cavalry at the siege of Veii.[27] The continuation and expansion of a group financially able to furnish their own mounts is implied in a speech of Cicero from 76. Cicero referred to C. Cluvius as an *eques* 'if you consider him from the census',[28] indicating that

23 Cf. Bourne (1952) 133.
24 *OLD* s.v. *excerpo*.
25 Livy 27.11.14 – *dena stipendia equis privatis facerent* 'they would make ten service years with private horses'.
26 Plutarch, *Pomp.* 22.4–6. Plutarch places the review at the end of service, but in Livy the censors examine the *equites* after the main census, Livy 29.37.8, (implied) 27.6–11. There is enough ambiguity in Plutarch's description to interpret the passage as referring to a point when those to be discharged were being reviewed, rather than meaning that only those to be discharged gave a full account of their service.
27 Livy 5.7.13 – *tum primum equis suis merere equites coeperunt*.
28 Cicero, *Rosc. Com.* 42 – *quem si tu ex censu spectas, eques Romanus est*.

the two groups were not the same. Wiseman points out that Cicero is most likely using the term *eques Romanus* precisely.[29] Coupled with the evidence for cavalry at Veii, there is no reason not to apply this definition and the group to which it refers to the Middle Republic. Thus the punishment ordained in 209 was only for the surviving cavalry *equo publico*, reflecting the higher esteem in which they had been held.

Aulus Gellius expressed some surprise that Cato the Elder considered the loss of a public horse so disgraceful when it was not accompanied by a loss of equestrian status (which was determined by wealth).[30] Nonetheless, Cato's interpretation is supported by the fact that only those cavalry *equo publico* suffered this punishment following Cannae. Further, the men under consideration in 209 had already been sent to Sicily for the duration of the war.[31] A census had been completed between Cannae in 216 and 209, indicating that the punishment was intended as a disgrace by censors who felt that their predecessors had been too lenient.[32] Indeed, it is to be expected that those serving *equo publico* were those most likely to continue to hold political office. The removal of their previous service would have seriously delayed, if not obliterated, their chances of gaining a magistracy.

In terms of census records, the punishment meant that in effect the previous years these men had earned towards their *emerita stipendia* were expunged. It is not clear whether the deeds declared by *equites* at the review were recorded, but given the stigma against Cannae survivors in particular, their loss of status may have been made clear. However, that the men were once again required to provide ten years' service suggests that a mark to negate their previous service was made on the main census list, in addition to their being struck off the *equo publico* list. Again, long-term record keeping was required to calculate service and maintain an accurate record for use by the censors. For punishments such as this to be effective, or even realistically possible, the censors must have been able to review a citizen's service history. Moreover, the case of the *equites* in 209 once again demonstrates the interaction of lists derived from the census list with the main census, as can be seen with the *tabulae iuniorum*. The relative complexity of

---

29 Wiseman (1970) 74. Equestrians were registered in the census with the rest of their tribe. As Wiseman highlights ((1970) 68), for a man to be granted a public horse the censors needed to know that he met the property qualification, requiring all to be registered in the census proper. The *transvectio* as described by Plutarch was a separate review carried out by the censors relating specifically to a designated role, much as was the senate roll review. However, the conclusions drawn from the *transvectio* regarding the use of lists by the censors are still valid for this discussion. The list of those *equo publico* was ultimately derived from the census and interacted with it much in the same way as the *tabulae iuniorum*.

30 Gellius, *NA* 6.22.3 citing Cato.

31 Livy 23.25.7.

32 Livy 24.18.

Rome's military bureaucracy with the census as its linchpin is further revealed.

Finally, the few instances where ancient writers provide an insight into individual records support some form of military record in the census. In 186 the senate decreed that the man who uncovered the Bacchanalia scandal was to be treated as though his military service was complete.[33] Without a permanent record of this grant easily accessible to both censors and consuls holding the *dilectus*, this reward would have been essentially meaningless. The obvious place for this is on the census record. As Livy uses the term *emerita stipendia*, it is not unreasonable to conclude that the same entry was made for this individual as for those who reached it by the traditional method. Thus, it appears that such records were kept on the census and, from that, included in documents derived from the census declarations.

The issue of how the records were kept, and by whom, will be addressed in detail in V and VI, but a brief mention needs to be made here. Particularly in the 214 example, but also generally, there is an issue of what occurred during the years between censuses. In 214, the previous census predated the Hannibalic War. That the censors were nonetheless able to punish those who had not served suggests that the records were in some way updated in the intervening years. Several modern scholars have reached this conclusion.[34] If this emendation occurred, it demonstrates that the military bureaucracy possible in Mid-Republican Rome was a great deal more complicated than is commonly allowed. Moreover, the censors were able to establish not only how many years had been served but when.[35] Importantly here, it indicates a record with more detail than just the number of years served. Together with the content of the census declarations, legion lists and previous census records could be utilised by the censors to create a quite detailed record of the military service history of each Roman citizen. The census formed the central pillar of this; it was the document from which other records held by the censors were ultimately derived.

ii Census registration on campaign

The previous section has demonstrated that Rome had a military bureaucracy able to track each man but was reliant on the census as its primary record. This raises a difficult issue: were citizens on campaign included in the census? The traditional position, particularly espoused by Brunt,[36] is

---

33 Livy 39.19.4.
34 Bourne (1952) 133, 135; Suolahti (1963) 45; Toynbee (1965a) 448.
35 See VI:iii.
36 Brunt (1971) 70–1; see also, e.g., Frank (1924) 330–1.

that legions outside Italy were not included in the census; the missions sent out by the censors in 204 were extraordinary.[37] Brunt bases his calculations of Roman and Italian population size and his demographic reconstructions on this argument.[38] This section aims to demonstrate that while 204 was an extraordinary case, roughly 50% of those on campaign were nonetheless registered at Rome. Coupled with legion lists, Roman military bureaucracy could still function with several thousand fighting men abroad.

First, however, the usual census process must be established. The census took place in Rome; there is no evidence for a local census accurate man-power records would be in the Middle Republic.[39] Modern scholars largely agree that families were registered by their *paterfamilias*; that is, only men *sui iuris* had to present themselves to the censors.[40] Livy, Dionysius and Gellius all provide evidence of the *patresfamilias'* role at the census.[41] Livy details the 169 census, when soldiers absent from the Macedonian legions were of particular concern. The censors requested that the men with *potestas* over these soldiers present themselves. This could be seen as evidence of a second-century local census such as that in the *tabula Heracleensis*, but need not be. Rather, it is a requirement that the *patresfamilias* speak directly to the censors rather than an official assisting the censors (see VI:iii). Likewise, the formula given with the census figure, *censa sunt civium capita*, 'the heads of the citizens were counted', might imply that every man over 17 was required to attend the censors in order to have his head counted in person.[42] However, this formula refers to the published census figure, which, while it included all these men, did not require them all to be presented in person. Rather, this number was generated from the returns provided by those *sui iuris*. The *tabulae iuniorum* already discussed demonstrate that the censors were capable of creating new lists from the census. Creating a total of all those over 17 need not have been any more complicated than generating the *tabulae iuniorum*.[43]

---

37 Livy 29.37.5; Suolahti (1963) 34.

38 Brunt (1971) 61–83.

39 The twelve Latin communities forced to send returns to Rome in 204 are not evidence for a local census. They were not Roman citizens; Rome was attempting to stamp her authority on recalcitrant colonies by forcing the essentially foreign *poleis* to conform to Rome's institutions. Livy 29.15.9–10. Cf. Toynbee (1965a) 221.

40 E.g., Toynbee (1965a) 445; Brunt (1971) 15; Nicolet (1980) 68; Lintott (1999) 117–8; Briscoe (2012) 434. It is not necessary here to discuss whether those declaring to the censors included widows and orphans, who were *sui iuris*, or whether their declarations were made by guardians. Neither widows nor orphans (by definition under 17) could serve in the army.

41 Livy 43.14.7–10; Dion. Hal., *Ant. Rom.* 4.14.6; Gellius, *NA* 4.20.3 (indirectly by mentioning the declaration of a wife); Cicero, *Leg.* 3.7.

42 E.g., Livy 35.9.2, 38.36.10; [Livy], *Periochae* 14.5, 20.15, 48.2.

43 Several modern scholars agree that lists for different needs were generated from the census declarations: e.g., Bourne (1952) 133; Suolahti (1963) 44; Toynbee (1965a) 445, 453–5; Hin (2008) 214.

During the Middle Republic the census declaration was given by the *paterfamilias*. Only he attended the censors, not those *in potestate*.

Before continuing, a brief note is appropriate concerning changes made to the census process by the censors in 169. As mentioned previously (I:iv), the censors in this year instituted a new question under oath, requiring any male citizen under 46 who had yet to serve to swear that he would attend the *dilectus*.[44] There can be little doubt that this oath had to be sworn in person. Therefore, the oath's establishment required all those under 46 who had not served, whether *sui iuris* or *in potestate*, to attend the censors. However, this is not of concern here, as those on campaign, by virtue of their service, were not part of this group.

Having established the usual convention for census declarations, the method for the army on campaign can now be examined. The best evidence is provided by the census proceedings of 204. In this year, the censors sent commissioners to the provinces to establish the numbers serving with the legions. Livy is explicit that this number was then included in the census total.[45] Toynbee argues this was the first time that soldiers on active duty were included in the census.[46] On the other hand, he himself points out that prior to the Hannibalic War, campaigns rarely lasted longer than seven months, meaning that all soldiers were home at some point during the 18-month censorship.[47] On this basis, the inclusion of men on active service was new and unusual precisely because it had not been required previously. Only the Hannibalic War's extraordinary duration precipitated this measure, because Rome was losing track of her manpower.

However, De Ligt has demonstrated that campaigns ending in December or January rather than October, and even winter garrisons, were not uncommon from the second half of the fourth century.[48] Short campaigns do not entirely rule out Toynbee's hypothesis, as the censorship period allowed men's inclusion. However, winter garrisoning (attested as early as 342[49]) does not allow for later registration. Nonetheless, it is likely that underregistration like that seen in the Hannibalic War was unprecedented, as the number on campaign in the fourth and third centuries was much lower.

Senatorial concern in 204 over under-registration is demonstrated by the unusually low 209 census figure of 137,108.[50] This figure has been

---

44 Livy 43.14.5–6.
45 Livy 29.37.5–6 – *lustrum conditum serius quia per provincias dismiserunt censores, ut civium Romanorum in exercitibus quantus ubique esset referretur numerus. censa cum iis ducenta quattuordecim milia hominum* (emphasis added).
46 Toynbee (1965a) 449.
47 Toynbee (1965a) 449.
48 De Ligt (2007) 119–20.
49 Livy 7.38.4.
50 Livy 27.36.6–7.

amended by scholars attempting to recreate Roman and Italian demography,[51] but Frank argues for its acceptance as transmitted.[52] It is not a realistic estimate of the Roman male adult population; it is demographically impossible for the population to fluctuate so violently over such a short period. Rather, the 209 census figure reflects what the censors were able to achieve during their magistracy. Thus it seems that Suolahti is correct to conclude that the missions of 204 were an exceptional case in a time of war.[53]

However, what made the mission of 204 extraordinary requires further consideration. It is possible that 204 was unusual because commissioners were sent to the legions rather than because totals were obtained from the legions. The frequent mention of dispatches to and from legion commanders opens the possibility that a list of living soldiers (more accurate than the legion lists held by the censors) could have been sent to Rome.[54] This interpretation is made more likely by events following the Battle of Cannae in 216. Livy records that the surviving consul, Varro, joined his cavalry with most of the surviving (free) foot and sent a dispatch to Rome. On receipt of this, the senate was able to inform the citizens of the deaths at Cannae.[55] Specifically, Livy says that families were informed of *privatae clades* (private disasters). The verb used, *uolgo* ('I publish or make known'), does not indicate the exact mechanism used to inform the bereaved; a posted list or an announcement *in contione* are both plausible. Nonetheless, Livy's description indicates that Varro's letter contained names, not just numbers. A list of those with Varro (rather than the dead, fled and captured) is most probable on practical grounds.

Additionally, the later treatment of the Cannae survivors may lend further support to the case for a list sent by Varro. They were sent to Sicily as legions for the majority of the war's duration.[56] The men at Canusium are described by Livy as roughly a consular army (two legions) and may well have been sent to Sicily.[57] Their defeat remained a stigma on their records. A list sent by Varro may have acted as the record's foundation.

However, the situation in 216 was, like 204, highly unusual; the legions had essentially been wiped out, and Rome had lost the war with Carthage but for its obstinacy.[58] Further, Varro needed to provide only names, not the full declaration required by the censors. On balance, it is safest to

---

51 E.g., Brunt (1971) 13.
52 Frank (1924) 330.
53 Suolahti (1963) 34.
54 E.g., Polybius 10.19; Livy 22.11.6, 22.24.14, 22.30.7, 22.49.10, 22.56.1–2; Plutarch, *Fab.* 3.4, 7.4; Appian, *Hisp.* 9.49, *Syr.* 7.39.
55 Livy 22.56.1–4.
56 Livy 23.25.7.
57 Livy 22.54.6.
58 Patterson (1942) 322.

conclude that commanders did not regularly send home census returns for their legions. The censors could use the existing legion lists to mark a legitimate absence on the record. As will be seen, the census system worked at a level of precision at which this uncertainty was permissible. The mission of 204 was indeed extraordinary.

On the other hand, that census returns were not sent by generals in the field prior to 218 does not mean that this was also the case after the Hannibalic War. Rome was a state demonstrably concerned with tracking its manpower, and almost constantly at war. Following the change in the scope of Rome's wars brought about in the Second Punic War, continuing concern with accurate manpower records would be natural. Indeed, the cessation of citizen tax in 167 would suggest that the manpower element of the census became even more prominent.[59] Despite this, the sources suggest that censorial missions to legions in the provinces did not become the norm. The second-century census figures continued to fluctuate, with a particularly low count in 194;[60] a variety of factors influenced this, but a failure to include the legions abroad may well have contributed. This all suggests that the 204 missions were an extraordinary one-off.

Nonetheless, the lack of missions does not necessarily mean that no soldiers abroad were registered. As already established, the *paterfamilias* was responsible for declaring all those in his *potestas*. Using the same models generated in the previous chapter (I:iv), it is possible to estimate what proportion of soldiers had a *paterfamilias* to declare them to the censors. Again, the same caveats must be emphasised: the models are exactly that, giving a sense of scale rather than a precise answer.[61] It is, in any case, impossible to know from what proportions of different age categories each legion was composed; this most likely differed from year to year and legion to legion. A sense of scale is all that is possible.[62]

Once again, the Coale–Demeny[2] tables for Level 3 West Male and Level 6 West Male have been applied to Saller's population models for 'ordinary' (men aged 30 and women aged 20) and 'senatorial' (men aged 25 and women aged 15) marriage ages. The range of mortality at birth and different marriage ages will provide a 'high' and 'low' model, within which is the most likely proportion of men in the legions with a *paterfamilias*. The census figures for 234 and 164 have also been used again. In this case, there is no need to remove a percentage to represent the *proletarii*; a proportion rather than a number is required. Therefore, the numbers generated in Appendix I, Tables A.1.1–A.1.2 without the 10% representing the *proletarii* removed have been used. These have then been applied, in each age

59 Cicero, *Off*.2.22.76; Plutarch, *Aem.* 38.1. See IV.
60 204: 214,000 (Livy 29.37.6), 194: 143,704 (Livy 35.9.2), 188: 258,318 (Livy 38.36.10).
61 See I:iv, n. 107.
62 Scheidel (2009) 32–6 examined the number with living fathers only.

Table II.1 Percentage of men with a living *paterfamilias*

Model	15+ (17+)	15–44 (17–44)	15–19 (17–29)
Level 3 West Male 'ordinary' marriage	28.61 (25.54)	38.84 (35.65)	53.38 (50.96)
Level 3 West Male 'senatorial' marriage	33.54 (30.32)	45.06 (41.81)	60.82 (58.06)
Level 6 West Male 'senatorial' marriage	38.26 (34.99)	53.97 (50.70)	70.98 (68.19)

bracket, to calculate the average number of men over the age of 17 with a living father or grandfather. These calculations can be seen in Appendix II. The total proportions for each of these categories (Level 3 West Male 'ordinary' marriage, Level 3 West Male 'senatorial' marriage and Level 6 West Male 'senatorial' marriage) are summarised in Table II.1.

No distinction is made between 234 and 164, as using the same model results in the same proportions for both. Nor is a distinction made between the married and unmarried. While exemptions for marriage (see II:iv) affected the legion's character, resulting in a legion of predominantly men under 30, marital status does not affect mortality or the average life expectancy of an individual's father. Theoretically (which is all that can be produced here), the proportion of men with a living paterfamilias in the same age bracket would have been the same regardless of whether the men were married or not.

Proportions of men with a living *paterfamilias* have been given for those aged between 17 and 46, as that is the age range during which *iuniores* were nominally available for service. The proportion for those aged 17–29 has also been given, following Rosenstein's conclusion that the vast majority of those aged 30 and above were usually exempt.[63] The 15, 17 and 44 age brackets have been used as discussed in the previous chapter (I:iv). It can be seen that, depending on the model used, between 50.96% and 70.98% of men on campaign had a living *paterfamilias* to declare them in the census. In numerical terms, from a legion of 4500 men this means between 2293 and 3194 men (to the nearest whole man). Allowing for a few men aged over 30 in the legion, and for the fact that these numbers are extremes between which the most accurate estimate lies, it can be stated with reasonable certainty that on average more than 55% of men in the legions had a living *paterfamilias* and thus were declared in the census.

Even for those without a living *paterfamilias*, it was still possible to be declared at Rome whilst on campaign. Varro states that tribal officials could make a declaration on behalf of an absentee.[64] Brunt argues that this

63 Rosenstein (2004) 85–9, see I:iv.
64 Varro, *Ling.* 6.86 – *curatores omnium tribuum, si quis pro se sive pro altero rationem dari volet.*

account is outdated, referring to a time when the community was small enough for the census and the *lustrum* to take place in one day. Consequently, it has little relevance to an 18-month censorship.[65] Such a conclusion is unnecessary. Varro mentions the possibility of a proxy in an almost casual way. Even if the exact procedure described is outdated for the second and first centuries, the proxy was still current and a detail with which Varro expected his reader to be familiar.

Cicero also provides evidence of acting for others. He promises to sort out the census return for the absent Atticus.[66] Interestingly, what Cicero specifically promises is to prevent Atticus being entered on the census as absent, *ne absens censeare*, 'lest you be counted as absent'. This suggests that the censors still entered individuals in the census even if they did not appear; presumably the previous rating was retained, with a penalty for failing to appear.[67] This further indicates that previous lists were consulted for the creation of the new. Cicero is ensuring that Atticus will be able to register at the end of the census period, allowing him to conduct his business and be given an accurate rating.[68] Cicero does not here register Atticus as a proxy, but he is able to intervene with the censors on Atticus' behalf. This suggests that action like that evinced by Varro was possible.

Cicero also demonstrates that Atticus' absence would have been noted. Suolahti argues that the new census list was based on the old. Thus, men were called in that order, by tribe, freedmen and then *equites*.[69] As Atticus was not a cavalryman (that is, a military cavalryman rather than of equestrian status in wealth terms), he would have been called in the tribal stage. Thus his absence, and by extension that of anyone included in the previous census, would have been noted. Cicero, like Varro, was writing in and concerning the first century, but his information tallies with that of the earlier Republic. The 169 example demonstrates that the censors had a list of those on campaign with which to compare the census declarations.[70] Coupled with the case of Atticus, this suggests that censors were able to spot absentees, account for those absent with the legions and retain their previous rating.

65 Brunt (1971) 536.

66 Cicero, *Att.* 1.18.8.

67 The harsh penalties of Livy 1.44.1 and Cicero, *Caec.* 34 are not attested as occurring, but the possible severity of the sanctions indicate the stigma of being absent.

68 Cicero, *Att.* 1.18.8 – *sub lustrum autem censeri germani negotiatoris est.*

69 Suolahti (1963) 37 using Dion. Hal., *Ant. Rom.* 5.75.3, 4.15.6, Livy 38.28.4, 38.36.5, 43.16.1. Using Polybius, Hill (1939) 357–62 has argued that there was no separate equestrian census during the Middle Republic. The equestrian census mentioned by Suolahti on the basis of Livy is in fact the review of public horse discussed previously. Hill's argument for a *census equester* in the Late Republic is not entirely convincing, as his reference may well refer to the same review.

70 Livy 43.14–5, see earlier.

In the case of soldiers *sui iuris* on campaign, it was unusual for many to be missed in the census, at least in theory. They could arrange registration by proxy. This, of course, does not mean that the arrangement was made. If enrolled in the year before a census was due, a *paterfamilias* might have made such an arrangement between his appearance on the Capitol and attendance at the first reassembly. However, for those enlisted shortly after a census, it may not have been a consideration, especially as the campaign's duration was unknown, as indeed was the gap between census periods. The legion lists should have allowed the inclusion of even those who did not make arrangements for registration. Additionally, tribal mechanisms may have made this registration by tribal official almost automatic for the absent. However, if the system worked this smoothly, a serious reconsideration of the demographic implications of the Mid-Republican census figures is required.[71]

It is not the purpose of this study to discuss demography in detail, but a few considerations need to be noted. Firstly, there is no direct evidence of ordinary Romans organising to be registered by proxy. This is probably a consequence of the general lack of evidence for the non-elite in this period. Varro's casual reference to such registration suggests that it was reasonably common, but the idea that soldiers on campaign engaged in it remains hypothetical.

Secondly, it is unclear whether those registered by proxy or accounted for by legion lists were included in the published census figure. Those *in potestate* were not physically present at the census but were nonetheless included in the head count, *capita censa sunt*; it is not a great leap to suppose that anyone registered by proxy would be counted likewise. However, the process may have differed for someone known only from a legion list and a previous declaration. The possibility of death on campaign (see III:ii) may have caused censors to err on the side of caution.

Thirdly, it is possible that individuals who had died on campaign were included in the census before their *paterfamilias* received the news. However, the census total would still be more accurate in this case than if everyone on campaign were excluded. Indeed, the requirement in 169 for the *paterfamilias* of men serving in Macedonia since 172 to personally attend the censors suggests that it was usual for those abroad to be registered.[72]

Bearing these considerations in mind, it must be conservatively concluded that approximately 50% of those on campaign – that is, those with a *paterfamilias* – were registered in the census. Nonetheless, this conclusion still has an impact on demographic studies. Taking Brunt's population

71 For example, Brunt's ((1971) 61–83) calculation of population by adding the approximate number of men in the legions to the census figures would be rendered wildly inaccurate.
72 Livy 43.14–5, see earlier.

calculations as an example (and withholding any other objections to his methods),[73] the number which he adds to each census figure to represent serving men must be halved to more accurately represent those not included in the census. The census figures of 209 and 204, and to some extent of 194, can still be explained in the same manner as advocated by Frank,[74] but the second-century returns require further consideration. There were, of course, more factors than service abroad affecting the census, but they need not be discussed here. It is enough to suggest that this interpretation of the census and military record keeping supports a 'middle count' demographic reconstruction.

This understanding of census taking must affect how the 204 census is viewed. If approximately 50% of serving soldiers were registered by their *paterfamilias*, then the numbers returned by the commissioners led to an overcount. Around 50% had already been included by the censors. However, given that the census figure for 204 as transmitted (214,000) appears to have been rounded and is possibly still lower than might be expected in comparison with later figures, several possibilities remain. Firstly, it may be that the censors were willing to include a possible overcount in the knowledge that, despite their diligence, they had likely not registered all citizens. Thus the overcount would help make up this shortfall. However, it is difficult to understand how the censors would be able to find these unregistered men for recruitment other than in an emergency *tumultus*. Secondly, it may be that Livy does not provide full details of the missions. The lists from the legions were not simply added to the total from Rome. Rather, the same process as in a usual census was undertaken, with a cross-check to prevent repetition. In this case, the census figure for 204 need not be scrutinised.

Thirdly, it may be that it was with this extraordinary census that cross-checking census lists with legion lists began. Visiting the legions was unusual, but it began a bureaucratic system which allowed future censors to generate a more accurate census. This raises the question of why the 204 censors did not simply use the legion lists in Rome to conduct their investigation. If 204 was the beginning of this cross-checking phenomenon, it is possible that legion-list copies were not yet kept in Rome. As has been seen already, the shorter third-century campaigns may have meant that absence from the census was so rare that such a record was deemed unnecessary. It was only with the new scale and scope of the Hannibalic War that such measures were required. This points to the development of Roman military bureaucracy toward a new complexity as the city was forced to adapt to the changing circumstances of war.

To conclude, the processes involved in census taking reveal that Rome was able to track her manpower to a reasonable degree without requiring

73 Brunt (1971) 61–83.
74 Frank (1924) 330–1.

fact-finding missions to the legions. The Hannibalic War occasioned an upheaval in Rome, and the use of extraordinary measures,[75] but the city soon adapted to cope with the changes in the scope and scale of warfare, with almost no change in the census process. It could be argued that it was at this point that registration by proxy was introduced, but there is no reason why it cannot belong to the third century or even earlier. More importantly, the combination of census records and legion lists generated at the *dilectus* allowed Rome to function at a relatively high bureaucratic level in order to keep accurate records. Even in the second century, at least half of those abroad on campaign were included in the census figure, and it is probable that this proportion was at times significantly higher.

iii Polybian manpower figures

Rome kept a reasonably accurate record of her manpower during the Middle Republic. However, in times of emergency it was possible to generate a more immediate picture of Rome's and her allies' military capability in a short period. In the second book of his *Histories*, Polybius provides a breakdown of the manpower resources available to Rome at the point of the 225 Gallic invasion. This breakdown has generated much scholarship on issues ranging from ancient Italian demography to the proper understanding of the Republican census figures;[76] little consensus has resulted. For this discussion, investigating the figures sheds light on the methods by which manpower was calculated and recorded both under duress and in more peaceful periods. Other attestations of Polybius' figures complicate the picture, but with full investigation broaden understanding. Viewed in line with near-contemporary census records, the breakdown provides a crucial insight into military administrative processes in the late third century, the census' weaknesses and limitations and the methods by which these shortcomings were circumvented.[77]

Polybius records the forces available to Rome against the Gauls in 225 as reported in καταγραφαί by the allies. These appear to have consisted of lists of men either currently serving or able to serve in the army. The totals given by Polybius are set out in his order in Table II.2. It is generally accepted that these originate from the καταγραφαί themselves, through the work of Fabius Pictor.[78] Fabius was a senator in 225 and active during the following Gallic War.[79] It is entirely plausible that he could have obtained access to these

75 Livy 29.37.5f.
76 See following references. Isayev (2017) 20 calls the figures the most useful record for Roman demographic calculations.
77 This discussion will necessarily touch on issues of Italian demography, but any attempt to add to the debate is beyond the scope of this monograph.
78 Walbank (1957) 184, 196; *FRHist I* 175–6.
79 Orosius 4.13.6.

Table II.2 Polybius' 225 manpower figures (from Polybius 2.24)

	Region	Foot	Horse
Levied	4 legions of Roman citizens	4 × 5200	4 × 300
	Allied contingent	30,000	2000
	Sabines and Etruscans	50,000+	4000
	Umbrians and Sarsinates	20,000	
	Veneti and Cenomani	20,000	
	Roman Reserve	20,000	1500
	Allied contingent	30,000	2000
Able to bear arms	Latins	80,000	5000
	Samnites	70,000	7000
	Iapygians and Messapians	50,000	16,000
	Lucanians	30,000	3000
	Marsi, Marrucini, Frentani and Vestini	20,000	4000
Levied	1 legion in Sicily and Tarentum	2 × 4200	2 × 200
Able to bear arms	Romans and Campanians	250,000	23,000
TOTAL	Romans and allies levied and able to bear arms	700,000+	70,000

καταγραφαί, whether they were held in an official 'archive' or copied into private memoirs (see V:iii), when he came to write. Indeed, the unusual geographic layout of the figures, with forces listed north to south following the active Roman legions, points to an emergency reaction to a northern incursion rather than invented or remembered figures. The very trouble of interpreting such an odd set of figures points to their reliability.

Polybius himself does not explicitly state that he used Fabius here, although earlier he mentions Fabius as an authority for the period.[80] An alternative source is a consul of 225, one L. Aemilius. Polybius does not provide a cognomen, but Walbank and Broughton conclude that this man was L. Aemilius L. f. Cn. n. Papus.[81] However, of all the evidence produced, only Appian and the *Fasti Capitolini* support Papus.[82] As an Augustan creation, the *Fasti*'s accuracy is uncertain and in places entirely incorrect. Taylor considers Livy to have recorded the true Republican tradition, not the *Fasti Capitolini*,[83] but the confusion in sources probably using Livy suggests that Livy did not provide a cognomen. Pliny gives Paulus, and Orosius Catulus.[84] As no other Aemilii Catuli are attested, Orosius can be ignored, but Paulus remains a possibility. This is particularly tempting as an Aemilius Paulus would be an ancestor of Polybius'

80 Polybius 1.14.
81 Walbank (1957) 196; Broughton (1951) 230.
82 *Fast. Cap.* XVIIa 529; *Fast. Trium.* XV 529; Appian, *Gall.* 1.2.
83 Taylor (1946) 8, (1951) 78.
84 Pliny, *HN* 3.138; Orosius 4.13.5.

friend Scipio Aemilianus.[85] It is possible that Aemilius recorded the returns in his magisterial *commentarius*; Polybius' close ties to the family may have provided access to such a record, a source possibly independent of Fabius. However, on balance the correlation between the consular and triumphal *fasti* suggests that in this instance Papus is the correct attribution. Nonetheless, it remains possible that Polybius was informed by an Aemilian *commentarius*. It cannot be ruled out that Fabius obtained his information from the Aemilii Papi as well, but his personal relationship with this *gens* is unclear. Such a source for the Polybian figures must remain hypothesis, but it adds to the potential reliability of Polybius' record.

Polybius is not the only source for these manpower figures. Diodorus, Pliny the Elder, Orosius, Eutropius and [Livy]'s *Periochae* all report the totals, with some variations (Table II.3). Orosius and Eutropius, both giving a total of 800,000, explicitly attribute the number to Fabius.[86] The *Periochae* also gives 800,000, although without citation.[87] However, it has been suggested that in *sui Latinique nominis DCCC milia armatorum habuisse dicit*, Fabius is the lost subject of *dicit*, giving the entry more grammatical and syntactical sense.[88] Diodorus and Pliny seem to follow Polybius, with 770,000 and 780,000, respectively.[89] Modern scholars generally consider these all to be replications of the same original.[90] It is not unreasonable to consider Fabius the ultimate source. However, the variation in these figures points to a more complicated historiographical inheritance than usually allowed, and requires further comment.

Eutropius, Orosius and, obviously, [Livy]'s *Periochae* were all heavily reliant on Livy for their work. That the 800,000 manpower figure was transmitted to them through Livy's lost books is almost certain, although an epitome may have been intermediary. Cornell *et al.* demonstrate that Orosius' 'preoccupation with cataloguing misfortune' has led to the reproduction of many numerical fragments of Livy in Orosius.[91] There is no reason to assume a lesser level of care in this case. (The foot and horse breakdown provided by Orosius will be discussed later.) This suggests that Orosius and – as they record the same figure – by extension Eutropius and the *Periochae* faithfully reproduce Livian figures.[92] The direct assertion of

---

85 Polybius 2.23.5; Pliny, *HN* 3.138.
86 Orosius 4.13.6; Eutropius 3.5.
87 [Livy], *Periochae* 20.
88 *FRHist II* 97, *FRHist III* 36–7 – '[Fabius] says that they and those of the Latin name had 800,000 soldiers in arms' (author's translation).
89 Diod. Sic. 25.13; Pliny, *HN* 3.138.
90 Walbank (1957) 199; *FRHist III* 37.
91 *FRHist I* 101.
92 The *Periochae* is anachronistic when it describes the 800,000 as *sui Latinique nominis*. However, misunderstanding the status of what the figure represents does not prevent the number itself from having been correctly transmitted.

Table II.3 Other 225 manpower figures

	Foot	Horse	Total
Diodorus Siculus	700,000	70,000	–
Pliny the Elder	700,000	80,000	–
Eutropius	–	–	800,000
[Livy], *Periochae*	–	–	800,000
Orosius	348,000 (Romans)	26,600 (Romans)	800,000

Fabius as the source by Eutropius and Orosius seems to indicate that Livy transmitted the manpower figures directly from Fabius. It is probable that Livy had read Polybius' version; he explicitly mentions Polybius in his work, and books 21 and 22 correspond to the extant books of Polybius.[93] However, Eutropius and Orosius should not be ignored. It seems that Livy took the figures directly from Fabius, not Polybius. Whether Fabius himself gave a figure of 800,000 or Livy rounded the given figure cannot be determined.

The figure of 770,000 given by Diodorus probably comes from Polybius or Fabius.[94] Both are plausible sources for Greek-speaking Diodorus. That the only explicit use of Fabius gives 800,000 might suggest Fabius over Polybius, but as these assertions came through Livy, it cannot be certain. As with those reliant on Livy, the historiographical sequence is unclear. The figure provided by Pliny the Elder is more interesting: 700,000 foot and 80,000 horse is generally considered a textual corruption from the 700,000 and 70,000 found in Polybius.[95] However, as Polybius states that there are more than these figures, it is possible that Pliny's figure is a more specific approximation. This suggests that Pliny was using an alternative to Polybius, and that this source was more exact. Polybius' own overall totals are a rounded figure based on the addition of rounded figures which appear to err on the lower side. While this does not aid in identifying Pliny's source, it demonstrates, along with the foregoing discussion, that the extant figures were not the only tradition in antiquity. These alternative figures were most likely unrounded versions of Polybius' and do not call them into question. On the other hand, they do suggest that more attention should be paid to the other extant figures rather than dismissing them as repetitions.

Having established the relative reliability of the extant figures as a representation of the original καταγραφαί, the meaning of the figures themselves must be examined. A common, although not universal,[96] complaint is

93 Livy 30.45.5, 33.10.10, 34.50.6, 36.19.11, 39.52.1, 45.44.19.
94 Walbank (1957) 199. There are no MSS deviations or modern emendations of this figure.
95 Walbank (1957) 199.
96 Lo Cascio (1999) 167, (2001) 131; Rosenstein (2002) 177–8; *FRHist III* 37.

that Polybius (or Fabius) has miscalculated his overall total by including those already levied twice.[97] The six active legions and reserve force are already covered by the total of Romans and Campanians, giving a citizen total of 325,300 rather than 273,000. This double-count assumption is based on the notion that Polybius' figures are essentially census returns. The figure of 273,000 fits the sequence of extant census figures, where 325,000 does not. The census return of 234 in particular – 270,212 – is cited as an indication of Polybius' inaccuracy.[98] For a 'low-count' understanding of Roman and Italian demography, Polybius' total is simply too high. There are several serious problems with this interpretation of the Polybian figures, not least that low-count demography requires the manipulation of the figures in this way. It is better to explore the options before assuming that a mistake has been made.

Who was included in the census as opposed to the Polybian figures is significant. Putting aside the problems of age groups, centurial classes and under-registration, there is a more fundamental problem. While citizen and allied contributions to the legions and reserve force are separated out, the rest of the καταγραφαί refer to geographical area rather than citizen status. For example, as Table II.2 shows, the Sabines and Etruscans were included as one entry. As Lo Cascio and Baronowski have noted, the Sabines were Roman citizens in 225, as were some Etruscans.[99] For the Etruscans the number of citizens was likely in the low thousands, as Caere was the only town partly enfranchised.[100] As *cives sine suffragio* they may have been infrequently registered in the census; coupled with Polybius' approximation of 'more than' for the number of Sabines and Etruscans, they have little effect on the calculations. Other fully enfranchised Etruscans had been placed in Roman tribes in 393 and 389, thus they will have been included in the figure of Romans and Campanians.[101] However, the Sabines will also have been ordinarily included in the census. If Lo Cascio is followed in considering the relative ratio of Etruscan to Sabine population to be about 3:2, around 21,600 (20,000 foot and 1600 horse) must be added to the low-count figure of 273,000 to give the actual corresponding census figure of 294,600. (It must be noted that this includes the assumption that foot and horse contingents were also equal in this ratio. Although with no evidentiary basis, this provides a starting point for these calculations.)

A total of 294,600 is not in itself an entirely implausible census figure for 225, but it does remove the appealing correlation with the 234 return. On

---

97 Walbank (1957) 198; Toynbee (1965a) 498; Brunt (1971) 45–7; Shochat (1980) 15–6, 35; Scheidel (2004) 3. De Ligt (2004) 735–6 finds neither argument entirely convincing.

98 [Livy], *Periochae* 20.15.

99 Lo Cascio (1999) 168; Baronowski (1993) 190.

100 Livy 7.20; Harris (1971) 45–7.

101 Livy 5.30, 6.3.4.

the other hand, if Brunt's estimate of 10% under-registration is correct, 294,600 would be almost exactly what might be expected during an emergency levy, where fewer would be missed.[102] However, this theory's requirement that the Polybian figures be seen as census returns means that an explanation using an emergency levy cannot be accepted. Toynbee gets around this problem entirely by considering Sabines a mistranscription of *Sapinia*, but his theory has not gained a great deal of traction.[103] Rather, the double-count interpretation loses ground.

A fundamental but unspoken assumption of the double-count interpretation is that it was usual for those on service to be included in the census in this period. As shown already (II:ii), the author does not contest that some of the serving were included in the census, but those arguing for the double-count almost always do so.[104] It is possible to argue that in 225 Rome was more mobilised than usual, in preparation for a Carthaginian rather than a Gallic war,[105] which may explain the minimal 'census' increase from 234. However, it cannot be had both ways. This flaw requires a reworking of many of the demographic assumptions and theories of these scholars. Polybius cannot be accused of double-counting if the serving were not regularly included in the census.

At this point it is worth examining the figures transmitted by Orosius. Along with an overall total of 800,000, the manuscripts all give citizen numbers as 348,000 foot and 26,600 horse, for a total of 374,600. The foot number is considered a transcription error, and can be changed with equal palaeographical plausibility to 299,200 or 248,200. The horse figure can be adapted to 26,100.[106] Both of these numbers fit a theory concerning the interpretation of the Polybian figures.[107] However, the very fact that both then fit so neatly with different interpretations raises questions of reliability. If both are equally plausible, they are equally implausible. There appears to be an issue of finding what one is looking for. Further, Shochat considers the alteration of the horse figure arbitrary and not justified.[108] It is difficult to imagine that the same might not also be true of the foot. To the author's knowledge, there has been no attempt to make sense of the manuscript figures. What follows, although imperfect, is an attempt at interpretation using the original figures.

First, Polybius. On the basis that he has not double-counted, Polybius provides a citizen total of 325,300. To this can be added the approximation of Sabine citizens calculated previously (21,600), giving a citizen total of 346,900+. Admittedly this is an approximation, but it is probably slightly low and is not too far from Orosius' total of 374,600, lending it some

102 Brunt (1971) 35.
103 Toynbee (1965a) 485.
104 Frank (1924) 330; Brunt (1971) 36–7.
105 Erdkamp (2009) 508.
106 Brunt (1971) 46.
107 248,000 Brunt (1971) 47; 299,200 Shochat (1980) 33.
108 Shochat (1980) 33.

Table II.4 Comparison of Roman citizen numbers

	Foot	Horse	Total
Orosius	348,000	26,600	374,600
From Polybius including Sabines	319,200+	27,700	346,900+
From Polybius including Sabines and Etruscans	349,200+	30,100	379,300+

plausibility. When this is broken down into foot and horse, these figures look less plausible, with 319,200 foot and 27,700 horse. This assumes the same ratio of foot to horse in both forces. It appears that the numbers derived from Polybius differ from those provided by Orosius to an extent that Orosius' figures must be considered incorrect for whatever reason (Table II.4).

However, there is a final possibility: only a few thousand Etruscans sent south to defend Roman territory are included in the figure of Etruscans and Sabines. If the whole figure (54,000), rather than an estimate of Sabines, is added to Polybius' figure, the overall Polybian total (now 379,300) is 4700 higher than the Orosian total – that is, roughly a Roman legion.[109] Broken down, this gives 1200 foot and 3500 horse. This is a very high proportion of cavalry, even for the allies, who regularly supplied more than the Romans, but it does not rule out this interpretation. Polybius states that there were 'more than 50,000' Etruscan and Sabine foot levied. What 'more than' equates to is unclear, but it may well make the ratio more recognisable, although still with a large cavalry contingent. As Polybius describes this part of the list as those defending Roman territory,[110] it is not implausible that the Etruscans sent only one 'legion' south to the praetor's command when the enemy approached from the north. Indeed, it helps explain why such an apparently huge force was entrusted to a praetor.[111] Moreover, this 'legion' was raised by the Etruscans apparently on their own initiative. There is no reason to assume that an Etruscan levy, especially an emergency levy, would work to the same totals and proportions as the Roman levy. It will be shown later that, in the context of the emergency levy, such a high number of Sabines is not implausible, nor is a report of so few Etruscans. Thus, while largely hypothetical, the Orosian manuscript figures can be used in the interpretation of the Polybian manpower figures and the reconstruction of events surrounding the Gallic invasion. From this, it appears that Orosius (and so Livy) had access to an alternative breakdown or a more detailed account of the war. Careful use of this evidence can help shed further light on Polybius.

109 379,300 from 374,600.
110 Polybius 2.24.8–9.
111 Polybius 2.24.6; cf. Toynbee (1965a) 483. Alternatively, recent scholarship may render a praetor commanding a large force unproblematic, see n.138.

A further argument against the double-count interpretation comes from examining exactly what Polybius claims to record. Polybius states at the outset that the Romans asked for ἀπογραφὰς τῶν ἐν ταῖς ἡλικίας, 'lists of those in the [military] ages'.[112] Brunt argues that ἀπογραφαί is a term Polybius uses to refer to the census, but his only evidence is this precise instance.[113] 'Reports' or 'lists' is a better translation, as this discussion demonstrates. Polybius' list of forces actually begins with those brought together (συνήχθησαν) – that is, those already levied[114] – and listing camps, κατέγραφον στρατόπεδα.[115] Once these men have been enumerated, he gives the καταγραφαί reports. The inclusion of the legions in Sicily and Tarentum in this part of the list is probably due to geographical expediency. Those as yet unlevied in central Italy would reach the Gallic threat before the distant standing legions. This in itself helps reveal the sense of Polybius' ordering. As Lo Cascio points out, the forces are listed in logical order for a defence strategy, with decreasing immediate availability down the list.[116] Thus not all the returns represent the same things. As made clear in Table II.2, Polybius includes lists of both the levied (συνήχθησαν) and those able to bear arms (implicit in the language).

However, this on its own does not rule out a double-count. It is necessary to establish how these figures were obtained. It has already been demonstrated that they cannot be considered census returns. Pliny describes the emergency as due to *gallico tumultu*, and Polybius refers to the terror of the old invasion.[117] It is clear that an emergency levy is taken across Rome and her allies. Rosenberger argues that the terror element should be ignored and the *tumultus Gallicus* seen as a purely pragmatic matter, but his explanation is not convincing. The very evidence he cites, the not-unproblematic Ceasarian *Lex Ursonensis*, singles out a *tumultus Gallicus* as the only occasion on which all recruitment exemptions are to be ignored.[118] Even if the fear element is mythic, it is in many senses academic. In 225 a quick response was required to face the threat; the speed of the attack is demonstrated by the failure to have men facing the incursion into Etruria.[119] The scale of the response need not be seen as excessive.

A *tumultus* was not unprecedented, providing an insight into how the emergency levy was conducted. Previous occasions demonstrate

---

112 Polybius 2.23.9.
113 Brunt (1971) 638.
114 Polybius 2.24.7.
115 Polybius 2.23.8.
116 Lo Cascio (1999) 167; (2001) 129.
117 Polybius 2.23.7; Plin., *HN* 3.138. Cf. Golden (2013) 57.
118 Rosenberger (2003) 367 with *CIL* II.5439 I 3.31–2 – *nisi tumultus Italici Galliciue causa.* On the issues with the inscription, see Hardy (1912) 10–1; Brunt (1971) 520; Crawford *et al.* (1996) 395.
119 Polybius 2.23.4, 2.25.1.

that the military oath was given to all men of military age without exception.[120] The express use of *omnes iuniores* indicates this included the *proletarii*, not just the *assidui*. Polybius does not himself use the term *iuniores* or a Greek equivalent, instead referring to those 'in the ages'.[121] However, Hin has demonstrated that the use of this term in Roman historiography refers to *iuniores*, including its use by Polybius elsewhere in his work.[122] Frank argues that *seniores* must be included, as they could stand garrison duty in Rome.[123] While this interpretation is plausible, the inclusion of the *seniores* is usually explicit when it occurs.[124] However, it would be foolish to imagine that *seniores* would not fulfil this role if necessary. Several examples demonstrate citizens coming to arms at an alarm with no levy at all.[125]

There remains the problem of why Polybius' record of *iuniores* is higher than the total census figure for 234. The *tumultus* method is key here. It appears that the *tumultus* was undertaken in a similar way to the 212 emergency levy detailed by Livy.[126] As discussed previously (II:ii), the 212 levy involved the enlistment of everyone suitably fit whom the commissioners came across, including those under 17. This has several implications: firstly, unlike the census, there could be no under-registration; and secondly, in 225's climate of imminent danger it is unlikely that those conducting the *tumultus* were overly concerned with the ages of those they enrolled. Using the proportions for the 234 census figure generated in Appendix I, Table A.1.1, a figure of 321,876 men can be generated by including *iuniores*, *seniores* and those aged 15–16 and assuming 10% under-registration. The possible mobilisation of all the *seniores* may be extreme, but the Gauls were not just another enemy to the Romans. The number can be made up to 325,000 if a small increase in population since 234 or a higher level of under-registration than the conservative 10% is allowed. Lo Cascio goes as far as to suggest that under-registration was so high that the entire figure was *iuniores*.[127] It is this number – not the *iuniores*, but all those liable to serve Rome – which Polybius records from the καταγραφαί of those able to serve. Thus it does not correspond with the census figures, but forms a separate, more accurate record of manpower in a seemingly desperate situation.

---

120 Livy 7.9.6, 7.11.6 *omnes iuniores sacramento adegit*, 7.28.3 *dilectus sine vacationibus habitus esset*.
121 Polybius 2.23.9 – ἐν ταῖς ἡλικίαις.
122 Hin (2008) 192–3; Polybius 6.19; Dion. Hal, *Ant. Rom.* 3.65.4, 4.15.46, 4.16.3, 5.75.4, 11.63.2.
123 Frank (1924) 332.
124 E.g., Livy 6.2.6, 6.6.14.
125 E.g., Livy 7.12.2.
126 Livy 25.5.6–9.
127 Lo Cascio (2001) 123–4.

Scheidel questions why Rome did not field such a number in the Hannibalic War if it had the manpower potential.[128] This is not an unreasonable question, especially as Polybius includes the figures for comparison with Hannibal's forces.[129] The inclusion of the *seniores* helps to explain it, but it can be answered by further examining the emergency levy's process. Although the oath was administered on a large scale, there is only one recorded example when all those levied at this initial stage were taken into the field, and then only for a single day.[130] More usually, a number of legions were created.[131] This suggests that the καταγραφαί of the able represent the number to whom the oath was administered. These men had not yet been recalled for the next stage of legion formation; they are the manpower potential of the territories, rather than those actually in the field. Indeed, Polybius refers to them as κατελέχθησαν, 'they were picked from'.[132] Building on the work of Erdkamp, Rosenstein has convincingly argued that ordinarily a very low percentage of *iuniores assidui* were on military service; the manpower requirements of subsistence farming were in part met with careful selection of *iuniores* whose labour could be lost without serious detriment to their families.[133] At the same time, manpower unavailability, not manpower shortage, meant that the massively increased military requirements of the Hannibalic War could not easily be met.[134] Thus, the able-bodied manpower of 225 was only the total potential; it could not have been realised without the collapse of the subsistence economy.

Those levied are a different category in Polybius' list. It appears that these are men fielded in defence. Polybius refers to the legions, στρατόπεδα, which literally translates as 'camps'.[135] It is worth returning to the problem of the Etruscans and Sabines here. Brunt highlights that it would have been extremely strange for the Etruscans to send their entire manpower south just as the Gauls entered their territory.[136] If the earlier interpretation of the Orosian figures is correct (the author wishes to emphasise the 'if'), the Etruscans sent only a single 'legion' south. As this leaves the rest of the Etruscan force unaccounted for, it is problematic. However, it is possible that Orosius (or Livy) misinterpreted the information he had regarding who should be considered citizens. The remaining figure should then be considered to be composed of the other Etruscans and Sabines who remained in

128 Scheidel (2004) 3.
129 Polybius 2.24.1, 2.24.17.
130 Livy 7.11.6.
131 Livy 7.9.6, 7.28.3.
132 Polybius 2.24.8, 2.24.14.
133 Erdkamp (1998) 255, 261, 265; Rosenstein (2002) 183–8, (2004) 85–90.
134 Rosenstein (2002) 185.
135 Polybius 2.24.8, 2.24.13.
136 Brunt (1971) 52.

Etruria, who were met by the advancing praetor. It is not unreasonable that the praetor led such a large force at this point, as one of the consuls was still returning from his initial posting with the legions, the other was stationed at another possible point of incursion and immediate action was required.[137] As the highest-ranking magistrate present in the area, the praetor naturally took command.[138]

Alternatively, it cannot be ruled out that the levy process or its terminology were not the same in all the territories. The terminology used to discuss allied levies is identical to that used for the Roman levy, but this does not mean that the processes were.[139] In discussing the recruitment of the Samnite Linen Legion in 293, Livy implies that this contrasted with the normal practice familiar to Romans and what was normal for the Samnites.[140] However, there is no explicit testimony of a normal levy for any of the allies. Those referred to by Polybius as στρατόπεδα may have included those in the field and the potential manpower as a single figure rather than the separation found for the Roman forces. The geographical nature of the returns suggests that the levy took place with some haste; such conflation from the northern states should not be surprising. On the other hand, the possibility of more than one tradition of surviving figures, already discussed, provides an alternative. If Aemilius did record the lists in his *commentarius*, he may well have not been as thorough and exact as the original returns. Conflated figures might be what should be expected if Polybius used this source over Fabius. In this case the exact number fielded is unrecoverable, but the forces under the praetor are reduced to a more acceptable level.

Overall, the Polybian manpower figures cannot be considered alongside the contemporary census figures. The conclusions to be drawn from this regarding military administration are interesting. The very necessity of requiring immediate reports from the allies demonstrates that any standing information was not adequate. That a similar process took place in Rome itself suggests that the same was true of Roman administration. The census figure as transmitted was not a standing manpower figure. This supports the conclusion that only approximately 50% of those on campaign would have been included. Hin has highlighted that this should be expected, as the

---

137 Polybius 2.23.5–6.
138 It may be questioned why a dictator was not appointed, as in other examples of *tumultus Gallicus*. Alternatively, recent interpretations regard the praetorship's origins as an office equal to and older than the consulship, a status still visible in the early third century at least. If this is correct, the prominent role of the praetor is not problematic at all, cf. Bergk (2011); Drogula (2015).
139 Cf. II:ii; e.g., Livy 21.17.3 (*scripta*), 32.8 (*dilectus, scribere*), although often the terminology is implied from the immediately preceding instructions for Roman citizen enrolment.
140 Livy 10.38.

census had more than one purpose, included more than just *iuniores assidui* and went five years between registrations.[141] She suggests that subsidiary lists were created each year for the individual purposes; in military terms, this was to account for those entering and leaving the *iuniores*.[142] However, the emergency measures taken in 225 suggest that it was not sufficient. Only by administering the oath to all eligible could a true reckoning of Roman manpower be made.

The geographical nature of the records indicates several things. At first glance it seems to emphasise the situation's emergency nature, as there was no time to centralise or organise by status. However, the geographical, rather than citizen-status, record had military benefit. As well as indicating manpower, it gave the senate a rough disposition. The benefit of this is demonstrated by the use of the Etruscan and Sabine levies.[143] While this may seem somewhat obvious and fundamental, such information was undoubtedly valuable. As already mentioned, the list's ordering reflects a defensive strategy based on just this. Further, the geographical separation even among Roman citizens (as Romans, Campanians and Sabines) indicates an ability to register men at a more regionalised level. While revealing nothing about a localised level, it does reflect a degree of military bureaucracy often overlooked in the modern scholarship. Military action on anything but the smallest scale requires high organisation, but evidence of this is usually scant, implicit or missing in the extant sources. The Polybian manpower figures open a small window into this military administration.

iv The census and military administration

It is now possible to add to the previous chapter's conclusions. Not only did Rome generate lists of those serving in each legion at the point of the *dilectus*, but this was only one part of a more complex bureaucracy. The continued smooth functioning of the census without declarations being sent from the legions demonstrates that legion lists were kept in Rome for consultation by the censors. At least half of those on campaign were included automatically; those who did not have a living *paterfamilias* could also be accommodated whilst absent by cross-referencing the previous census list with lists of active legions. While this may have been a labour-intensive process, the duration of the census indicates that, with a support staff (see VI), it was possible for different documents to be used together. In this way, the census and legion lists from the *dilectus* operated symbiotically. The census list was used to create the *tabulae iuniorum* that were used to generate legion lists at the

141 Hin (2008) 195–7.
142 Hin (2008) 214.
143 Polybius 2.24.6–7.

dilectus. In turn, these legion lists assisted the censors in the creation of the most accurate possible list of available and liable manpower for the next five years in the following census. The records' interaction allowed the circumvention, by and large, of the apparent problem of not being included in the register of citizen manpower due to absence caused by fulfilling citizen obligations. The operation and management of what was in Polybius' eyes Rome's key military advantage, her manpower,[144] was possible due to the relative complexity of this administrative structure.

On the other hand, the Polybian manpower figures reveal that there were limitations to this bureaucracy. To gain an accurate immediate picture of her manpower, Rome needed to physically enlist all the liable men; the census and derived *tabulae iuniorum* were not enough in an emergency. As the census was only taken every five years, and on occasion less frequently, this limitation should not be all that surprising. Rather, it is a mark of Rome's ability to deal with emergencies that the limitation could be circumvented and an accurate picture generated so swiftly. The ability of both Rome and its allies to perform the levy on such a scale in a remarkably short period of time, probably only a few weeks,[145] and transmit the information to a central organising hub indicates a degree of bureaucratic preparedness many would find surprising in what remained essentially an oversized *polis*.[146] The military administration, which should not be too separated from the state administration, could function at a remarkable speed and accuracy.

Alongside the *dilectus* and its legion lists, the census performed a key role in holding together the administration of Rome's armies and manpower in Rome. It has been established that the census was the centralised administrative point, to which declarations direct from the armies and their commanders were not added. To broaden the developing picture of Roman Republican military administration, it is necessary to examine the function administration had within the legions, and what form this took. Only by combining this element with the *dilectus* and the census can a full picture of the bureaucracy surrounding the management and functioning of the legions be understood.

144 Polybius 6 *passim*, cf. Pelling (2007) 247.
145 Before the consul could return from Sardinia, Polybius 2.23 5, 2.27.1.
146 As noted by Scheidel (2004) 6.

3   Recording men on campaign

The census may have functioned as the central aspect of Republican military administration, but it was not the only part. Records carried with the army on campaign were equally important to Rome's military success. For tactical reasons it was important for a general to know the number of men under his command. This chapter argues that the general not only knew this number but also had a full list of the men under his command which detailed their rank, included their previous service and accounted for their pay. Such a list was based on the legion list created at the *dilectus* and organised at the first reassembly, but contained extra detail in order to run the legion as smoothly and efficiently as possible. In order to demonstrate this, several aspects of administration on campaign will be examined. Firstly, the importance of accurate manpower figures will be underscored by examining food supply. Secondly, the casualty figures provided by ancient writers will be used to investigate whether and how armies counted their soldiers. This reveals how the general was able to maintain a working total of the men under his command. Both the dead and deserters were accounted for to some extent, and it was possible to send a list of the dead containing at least some identities to Rome. Thirdly, the role of the quaestor as paymaster will be discussed. This demonstrates the necessary complexity of paperwork required to ensure that each man received his proper pay. Next, the interaction of records kept on campaign with those in Rome will be examined through the *supplementa* enlisted for existing legions, and finally the *lustrum* conducted by a new commander upon his arrival in his province.

i   The importance of accurate figures

A brief foray into logistics demonstrates the importance of accurate manpower figures to successful army operation. Understanding Republican logistics is difficult at best, as it tends to be overlooked by the sources unless something had gone awry.[1] On a very basic level, in order to have enough

---

1 Erdkamp (1998) 46; Roth (1999) 3.

food to feed a legion, it was necessary to know how many men needed to be fed. Undersupply was potentially disastrous. However, the issue is more complex. In order to understand the intricacies of food supply, several aspects of supply method must be examined.[2]

Firstly, the movement of supplies with the legion. Seemingly the most obvious way to keep a legion supplied and independent, especially in hostile territory, was to carry food with the legion. However, any calculation of the amount of food required to feed 4500 men even for a short period quickly reveals that this was impossible. Using Polybius' two thirds of an Attic medimnus of wheat per month per man (generally accepted as four Roman *modii*[3]), approximately 1 kg of wheat per man per day was required.[4] For a legion of 4500 men this equates to 18,000 *modii* or approximately 135,000 kg of wheat per month. For a legion of 6000 men it increases to 24,000 *modii* or approximately 180,000 kg. This must be doubled to cover the allied contingent. Further, this does not include other personnel within the legion, including the quaestor and his staff. The men and animals of the baggage train also required feeding, as did the cavalry horses. Moreover, if Feig Vishnia is correct that the *lixae* who accompanied the legion were also fed by the legion, the number to be fed and thus the amount of grain required was again increased.[5] The number who made up this accompaniment is uncertain, but it may be as high as the number of the legion itself.

In addition, the higher the number to be fed, the greater the increase of men and draught animals required in order to transport their food and fodder. In a study on the early modern army, Perjes calculated that for an army of 90,000 to be self-sufficient for one month required a column of 200 km to carry provisions, made up of 11,000 carts, 22,000 drivers and helpers and 50,000–70,000 draught animals.[6] As Erdkamp points out, this was impractical to the point of rendering the army unable to manoeuvre and impossible even for an army half that size.[7] Erdkamp has demonstrated that even if the men carried a large proportion of their own rations, a mule train of at least several hundred mules would always have been necessary. Despite this, it would only have been possible to operate in this way for a few days.[8]

---

2 Much more detailed and comprehensive studies have been completed by Erdkamp (1998) and Roth (1999), on whose work much of this section is based.
3 Erdkamp (1998) 27; Gilliver (2001) 345. *Contra* (3 *modii*) Walbank (1957) 722; Duncan-Jones (1982) 146 with caveats.
4 Polybius 6.39.13; Garnsey & Saller (1987) 89.
5 Feig Vishnia (2002) 268–70 with Sallust, *Iug.* 44.5. Feig Vishnia argues that the *lixae* were separate from the merchants and small-time traders who followed legions. They were particularly involved with capturing slaves on campaign, as well as probably helping with forage.
6 Perjes (1970) 11.
7 Erdkamp (1998) 20.
8 Erdkamp (1998) 75. For example, Erdkamp calculates that 3500 mules would be required to carry food and fodder for ten days if the men carried eight days' worth of their own rations.

Thus a legion could not carry with it all the grain required for a month, let alone a whole campaign season. The legion would necessarily be reliant on either constant resupply or local forage. Both of these methods carried their own problems. Resupply required a magazine to be created close to the combat area. Ideally this would be in a fortified, allied or occupied city with easy access to supply routes to the legion and to the grain source. Factors such as vulnerability, sea conditions, availability of transport and harvest yields could all have an impact on supply effectiveness.[9] Similarly, living off forage had its own difficulties. In winter, all the seed grain had been sown, and remaining stocks held by locals would be dwindling as they ate them. A static campaign would quickly go through the local supply even at harvest time. Living off the land alone was only a plausible option for a moving campaign during the harvest period. Cato the Elder's self-sustaining Spanish campaign of 195 is an example of a commander successfully doing so. However, it was far from the norm, as the presence of contractors wishing to supply the army demonstrates.[10]

As the organiser of an army's food supply, the quaestor had to balance the advantages and disadvantages of all these supply methods. Each campaign would have different circumstances for the quaestor to manage, concerning not only the local area and time of year but also the yield of the previous harvest. Several modern scholars have concluded that the acquisition of supplies and the exact people involved would thus have been quite ad hoc, reacting to circumstances and the demands and aims of the campaign and commander.[11] The number to be fed was high, but the large difference in the amount of wheat required for 4500 versus 6000 men highlights that even a relatively small change in the strength of the legion could have potentially huge effects on demand. Erdkamp argues that commanders could choose to operate 'at the fringe of the logistically possible' to gain an advantage, although this increased the risk of encountering problems.[12] Such a strategy made the number of men to be fed even more crucial. A commander could gain a greater advantage if it was known that only 5500 men rather than 6000 needed to be fed. On the other hand, operating with such narrow margins meant that even a small undersupply could potentially starve the legion if another method could not quickly supplement supplies. In order to successfully supply the army at a level that met tactical requirements but did not endanger it through starvation, the quaestor needed an accurate record of the men present.

---

9 Cf. Vegetius, *Mil.* 4.39; Livy 23.48.10–49.3.
10 Livy 34.9.12 – '*bellum*' inquit '*se ipsum alet*'[...] *id erat forte tempus anni ut frumentum in areis Hispani haberent.*
11 Erdkamp (1998) 153; Gilliver (2001) 346.
12 Erdkamp (1998) 142.

Noncereal elements of the military diet only added to the complexity of the quaestor's work. Appian reports that the military diet included wine, salt, vinegar/sour wine, oil and meat in addition to wheat.[13] Aulus Gellius states that soldiers were not required to hand over any fruit foraged, suggesting that it supplemented the diet as and when it could be procured.[14] Moreover, Appian mentions measures taken when supplies failed.[15] Together, the evidence indicates that the use of local resources was particularly important for providing perishables to supplement the wheat ration. This necessarily resulted in a less regular, more varied supply, requiring regulation based on the number of men present. For example, if animals were butchered for meat, the quaestor needed to know how many people were to receive a meat ration, to ensure that the right number of animals was slaughtered. Any leftover meat would quickly have spoiled and been wasted, or required time to salt or smoke. Once again, it is clear that accurate manpower records were essential, as even a relatively small difference in the actual strength of the legion compared to its original or paper strength could have a large effect on how the army was to be supplied.

ii Counting the dead

As the previous chapter demonstrated, the legions were not directly registered in the census. That does not rule out manpower records within the legions. Scholars have questioned generals' awareness of the exact sizes of their forces, especially with the effects of deaths in battle and from disease.[16] However, the foregoing demonstrates that it would have been highly impractical for commanders to have little idea of the operational size of their forces. Certainly, Polybius states that generals changed their camp to fit arrivals and departures, indicating that at least a rough record of the numbers present was kept and regularly updated.[17] Moreover, the frequent citation of casualty figures, particularly by Livy, suggests that generals both could and did ascertain reasonably accurate operational figures. This section examines these casualty figures in order to argue that generals at the very least attempted to discover the number of dead and living from their legions. Enemy casualty figures (whether from Roman or Carthaginian sources) will not be discussed here; the purpose is to identify aspects of Roman military administration, not provide a more general survey of casualty figures in ancient histories.

---

13 Appian, *Hisp*. 54. The exact nature of vinegar/sour wine is debated in modern scholarship but is not of concern here. It is only significant that wheat was not the only foodstuff which needed to be acquired. Cf. e.g., Middleton (1983) 75; Erdkamp (1998) 34.

14 Gellius, *NA* 16.4.2.

15 Appian, *Hisp* 54.

16 Harris (1971) 68; Brunt (1971) 694–7.

17 Polybius 9.20.

The source and reliability of ancient casualty figures must first be examined. The origins of the figures helps to reveal the nature of the records kept by generals on campaign. The majority of Mid-Republican extant casualty figures are found in Livy.[18] Livy himself most frequently cites Valerius Antias as his source for these figures.[19] Consequently, the casualty figures have been condemned as incorrect by some modern scholars on the basis that, unless otherwise stated, they derive from Valerius' work.[20] It is therefore necessary to examine what can be gleaned about Valerius' work and methods in order to better understand the casualty figures before they can be used to draw conclusions about military administration.

Valerius was an immediate predecessor of Livy, writing around 80–60. His work, no longer extant, appears to have been the major work of Roman history before being superseded by Livy,[21] and in recent years it has been studied in some depth.[22] Rich in particular has highlighted that Valerius' work seems to have been substantially longer than that of his predecessors.[23] This generates questions around how his history was expanded and whence the material was obtained.[24] An obvious and immediate answer here is 'plausible invention', a facet of ancient historiography recognised and accepted by Thucydides in the fifth century with regard to speeches and practised (if somewhat hypocritically) by Polybius in the second.[25] Valerius is particularly accused of attempting to glorify his *gens* by embroidering or completely inventing the antics of his ancestors, as well as adding colour to his work with the inclusion of fabricated numerical details.[26]

---

18 Numbers are notorious for corruption in the manuscript tradition. However, the majority of the casualty figures for books 31–45 can be accepted as correctly transmitted, with only a few variant readings in the manuscripts (Briscoe (1973), (1981), (2008), (2012); Livy 34.41.10, 37.44.1, 41.18.13). Nevertheless, it is possible that there are cases where an early copyist made an error which has been transmitted in all the surviving manuscripts. This kind of error is unidentifiable, so the casualty figures must be warily accepted as transmitted.

19 Livy 3.5.12–13, 25.39.14, 16–17, 30.19.11, 30.29.5–7, 32.6.5–8, 33.10.8, 33.36.13–15, 34.10.1–2, 34.15.9, 36.19.11–12, 36.38.5–7, 37.60.4–6, 38.23.8, 38.55.5–9. Cf. Orosius 5.3.3, 5.16.1–4; [Livy], *Periochae* 67.

20 Briscoe (1973) 11; Laroche (1988) 771; Erdkamp (2006b) 166–8. However, others have argued that the casualty figures for 218–167 can be regarded as correct on the whole to varying degrees: Toynbee (1965b) 45; Rosenstein (2004) 109; Brunt (1971) 694.

21 Rich (2005) 138–40 with Asconius, *Corn.* 69C.

22 Laroche (1988); Rich (2005), (2013); Erdkamp (2006a), (2006b).

23 Rich (2013); *FRHist I* 297.

24 Rich (2005) 144, (2013); Erdkamp (2006a) 560.

25 Thuydides 1.20–2; despite his comments here, Polybius 2.56, 12 passim.

26 Toynbee (1965b) 38; Laroche (1988) 770–1 (tempered); Erdkamp (2006a) 556, 560; Rich (2013). The author has elsewhere argued that such an explanation of Valerius' narrative should not be accepted for the actions of Valerius Corvus in the First Samnite War; Pearson (2012).

These 'fabricated' details, runs the argument, were reproduced by Livy due to his lack of methodological rigour in composing his histories. There is no evidence that Livy examined any primary material himself, indicating that he instead relied on previous histories.[27] He follows a main narrative for sections, using others only to supplement details or note discrepancies.[28] In the 'late-annalistic battle scenes' identified by Erdkamp, Livy includes a great deal of detail about battles and their aftermath, frequently including casualty figures (usually for both sides), prisoners and captured military standards.[29] The order of this information is the same in nearly every 'late-annalistic battle scene'.[30] Erdkamp argues that Livy here followed the work of a late annalist, almost certainly Valerius Antias;[31] Valerius is most frequently cited in these 'late-annalistic' narratives. Thus, Livy incorporated Valerius' 'fabricated' figures without any attempt at verification. This explanation is accepted by many modern scholars, and consequently requires rejection of the casualty figures.

However, the issue cannot be so easily settled. There is little reason to question Erdkamp's conclusion regarding the origin of 'late-annalistic battle scenes', or Livy's general working method. Nonetheless, it does not follow that Livy unquestioningly followed his sources,[32] nor that Valerius was a serial inventor. It was pointed out as early as 1906 that Livy's frequent citations of Valerius point to dissension, not agreement.[33] Each mention of Valerius suggests that Livy is unwilling to cite the casualty figure on his own authority, indicating an awareness of possible inaccuracy or corruption. On the other hand, on several occasions Livy compares figures from several sources, an exercise in which Valerius is not always judged unfavourably.[34] The frequent citation of Valerius may even be Livy calling on a superior authority, if the conclusions to come on the origins of Valerius' figures are correct. Thus, while Livy may not have consulted primary material, he nonetheless was concerned about the accuracy, or at least plausibility, of the figures which he included in his work.

First-century readers and writers appear to have found the inclusion of casualty figures in histories realistic. The Greek historiographical tradition

---

27 Toynbee (1965b) 36–8. See Introduction.
28 Oakley (1997) 17; Erdkamp (2006a) 547; *FRHist III* 349.
29 Erdkamp (2006a) passim.
30 Laroche (1988) 760.
31 Erdkamp (2006a) 557–60, (2006b) 170; cf. Laroche (1988) 758.
32 Contra Erdkamp (2006a) 539 '...military aspects of battle, such as tactical considerations or terminology concerning units, did not matter and could either be discarded or copied verbatim'. Moreover, it must be emphasised that the accreditation of any passage to a specific but uncited writer remains hypothetical, however enticing the argument, cf. *FRHist I* 299–301.
33 Howard (1906) 162. E.g., Livy 3.5.12–13.
34 *FRHist* 25 F36 (Livy 33.10.8), F37 (Livy 33.30.6–11), F48 (Livy 38.23.8).

may have been influential here. Thucydides regularly included casualty figures in his battle reports; these are accepted as accurate by modern scholars, and were even used for an examination of average casualties in hoplite warfare.[35] Their accuracy is further supported by Thucydides' own hesitancy to include figures when he could not find a report he considered trustworthy.[36] As an active general, his judgement in these matters can be trusted. Thucydides seems to have established including casualty figures in battle reports as the norm; Xenophon, Polybius and Claudius Quadrigarius are just three who followed suit.[37] However, there is no guarantee that authors following Thucydides were as scrupulous as he. For example, Claudius' figures are questioned by Livy.[38] A reader's expectation of inclusion is also no indicator of reliability. On the other hand, the turmoil of the first century meant that many of those reading Valerius' and Livy's works had personal experience of military service and administration. Livy frequently gives casualty figures in terms of scale, such as 'many' or 'more than average', suggesting that he expected his reader to know what ordinary ratios for victorious and defeated armies were.[39] The expectation of accurate casualty figures could equally have come from this Roman experience as from the Greek historiographical tradition.

This expectation suggests that first-century readers believed writers could access information about battle casualties. This indicates that, at least in the first century, a method of tracking manpower within legions was exercised by generals.[40] The same was not necessarily true of the Middle Republic, but opens the possibility. Further, Livy is explicit on several occasions that reports, both from written letters and oral testimony, were announced *in contione* following their discussion in the senate.[41] The use of *volgata sunt* ('made know or published') in particular, in a context indicating usual practice, suggests that this was a common occurrence usually not reported by Livy.[42] Indeed, Livy never mentions the public announcement of a commander's and legions' deeds in the context of a triumph, but nonetheless he expects his reader to understand that they were known by the

---

35 E.g., Thucydides 1.63.3, 3.109.2, 4.44.6, 4.101.2, 5.74.2, 6.71.1, 8.25.3; Krentz (1985).
36 Thucydides 3.113.6, 5.68.
37 E.g., Xenophon, *Anab.* passim; Polybius 1.34, 3.84, 3.117, 10.32, 15.14, 18.27; Claudius Quadrigarius, *FRHist* 24 F63 (Livy 33.10.8), F64 (Livy 33.30.6–11), F67 (Livy 38.23.10).
38 *FRHist* 24 F67 (Livy 38.23.8).
39 E.g., Livy 10.35.2 Following a similar methodology as that used by Krentz (1985) for deaths in Greek hoplite battles, Rosenstein (2004) 109 concludes that Roman infantry battles had an average mortality of 8.8% for the first 32 years of the second century.
40 Polly Low (pers. comm.) points out that this expectation may have been conditioned by previous works and does not necessarily reflect administrative reality. However, as much of the readership of first-century histories was familiar with military procedure, acceptance of the figures indicates that possessing such knowledge was considered plausible.
41 Livy 23.29.17, 27.7.9, 30.17.3, 30.40.3, 32.31.6, 33.24.4, 33.25.8.
42 Livy 23.29.17 – *quae posteaquam litteris Scipionum Romae volgata sunt.*

people. However, this does not prove that casualty figures were announced alongside the report of action. Despite this, that Livy expected his readership to recognise a 'normal' casualty figure suggests that even if deaths were not always announced, they were reported frequently enough for such knowledge to be commonplace.

What, then, was the source of Valerius' (and so the majority of Livy's) casualty figures? Valerius appears to have been the first Roman historian to regularly include such numerical details in his work; earlier writers had only done so sporadically. However, 'plausible invention' does not have to be the explanation for his greater detail. Rich argues that Valerius was responsible for the classic structure of annalistic history found throughout Livy's extant work from book 21. This comprised: senatorial decisions concerning military recruitment and deployment, including a list of the active and reserve legions, at the beginning; events in the field in the middle; and omens and elections at the end. Rich argues that Valerius developed this formulaic, thematic and strictly chronological structure through archival research.[43] So influential was this framework that even when largely following another author, such as Polybius, Livy applied it.[44] This further demonstrates that Livy did not unthinkingly copy his sources. Most significantly here, it suggests that Valerius had access to records detailing campaigns of the Middle Republic.

Rich is careful to state that using an archive does not mean that Valerius did not also use 'plausible invention', and that separating the two is not simple. He points to the uncertainties with the legion lists from the beginning of each year.[45] While these lists are not problem-free, Rich falls into the trap he warns against here: it cannot be definitively stated that the legion lists were taken directly from a senatorial list, but by the same token it cannot be established that they were not. From a tactical and administrative standpoint, keeping a record of legions was a sensible procedure, especially as the senate decided their deployment each year. Their existence can be further surmised from the recruitment lists discussed previously (I:v, II:ii). Even if Valerius did not find lists like those given by Livy, the senatorial decrees on recruitment and deployment repeatedly mentioned by Livy would have made their compilation relatively easy for either Livy or Valerius.[46] Any problems with the lists should perhaps be attributed to confusion or lack of clarity on the part of the authors rather than fabrication.

In terms of casualty figures, similar conclusions can be drawn. Rich argues that casualty reports originated in commanders' dispatches, although

---

43 Rich (2005) 156; reiterated Rich (2013).
44 Rich (2005) 156–7. Cf., e.g., Polybius 3–4 with Livy 21–2.
45 Rich (2005) 160.
46 E.g., Livy 26.1.1–13.

they may have been subject to later inflation.[47] However, there is little reason why Valerius could not also have taken casualty figures from such records. Ancient authors provide ample evidence of written and oral reports between generals and the senate concerning their activities and the state of their legions.[48] The accounts of events such as battles are usually preceded or followed by a notice of the arrival of a commander's letter.[49] This suggests that the account itself could originate from the letter. It is highly plausible that such quantitative details made it into more formal records. If so, Valerius could have taken directly from the senatorial records casualty figures which themselves came directly from the letters of generals in the field.

It has been suggested that a particular style of Latin was used for military communications. If correct, this could aid in identifying the source of details such as casualty figures. A high frequency of the ablative absolute is argued to be indicative of military style; Julius Caesar's *commentarii* are considered archetypes of this genre.[50] Plautus' use of the same form indicates that it was a specific style of military report. The second-century playwright's work contains very little ablative absolute. This makes the report of ex-soldier Sosia on his master's martial achievements all the more marked. The speech begins with a string of three absolutes which set the tone.[51] Both Leeman and Adams conclude that this change in style is not a parody of tragic or epic style, but a deliberate use of military style.[52] As such reports were read to the people as well as the senate, Plautus' audience would be familiar with the style and understand Plautus' aim in placing it in Sosia's mouth. Caesar's style was not new, but a continuation of a much older style of writing.

However, construction of the Latin does not allow the identification of the military ablative absolute and thus the sources used by Livy/Valerius Antias. The majority of the reports which directly accompany the mention of a letter or envoy are given in the accusative-infinitive construction, as expected of a reported speech or letter.[53] On the positive side, Leeman

---

47 Rich (2005) 148; cf. Walsh (1994) 142.
48 Written: Polybius 10.19; Livy 22.11.6 *inter alia* 45.2.2; Plutarch, *Fab. Max.* 3.4, 7.4; Appian, *Hisp.* 9.49, *Syr.* 7.39. Oral: Polybius 10.19; Livy 22.49.10 *inter alia* 45.2.2, including 23.25.5 and 44.20.2–7 explicitly on forces; Plutarch, *Cato Mai.* 14.3, *Fab. Max.* 3.4, 16.7; Appian, *Pun.* 7.48.
49 On the senate meetings called to hear these accounts, see Ryan (1998) 19–20.
50 Leeman (1963) 176–7; Adams (2005) 73–5.
51 Plautus, *Amph.* 188–9 – *uitores uictis hostibus legiones reueniunt domum/ duello exstincto maxumo atque internectis hostibus.*
52 Leeman (1963) 176–7; Adams (2005) 73. It may be questioned, as Beard (2007) 203 does, why Cicero's dispatches to the senate do not follow this form. However, this probably more reflects Cicero's self-presentation than the nature of military reports.
53 Examples throughout Livy 22.11.6–44.16.2, see especially 23.21.2–3, 23.24.1, 23.24.1, 23.29.17, 23.34.10–12, 23.48.4.

highlights that the first book of Caesar's *Gallic Wars* is 32% indirect speech, a much higher proportion than found in other historians.[54] However, in all cases, it is impossible to be certain of the letter's original form from the construction of the text, let alone whether Livy or Valerius had access to it. When an ablative absolute is found in a battle account, it is impossible to judge whether the form reflects the original letter or Livy adopted the appropriate military style.

Despite this, that Livy mentions a letter on almost all occasions when he discusses issues abroad, either before or after the account, suggests that the account derived from a letter, whether directly or indirectly. While it is impossible to be sure whether Valerius, and thus Livy, used military dispatches as direct evidence, both the existence of a military style and the frequent appearance of letters and in-person reports demonstrates a wider, and very significant point: at whatever distance they were used by historians, frequent communications between Rome and the field were an ordinary part of Mid-Republican warfare.

Varro's letter in 216 following Cannae (II:ii) demonstrates that it was possible for commanders to send relatively detailed news to Rome by letter.[55] If Varro could do this in an emergency, there is no reason to believe that other generals did not do so routinely. Additionally, the majority of the fragments of Valerius Antias are concerned with numbers, either casualty figures or treasury deposits.[56] Pelikan Pittenger suggests that these quantitative details could originate from the context of triumphal debates.[57] Alternatively, if Valerius consulted the records directly and is giving numbers connected to the treasury, it suggests that he used some kind of treasury record, perhaps similar to those reported by Pliny the Elder.[58] Thus these numbers need not be suspect, and have been accepted by modern scholars without sceptical comment. It is therefore plausible that casualty figures can be similarly attributed to a roll from the records. There is no reason whatsoever to believe that the figures are a wholesale invention of Valerius.

A further scepticism from modern scholars is the assumed difficulty for generals of generating these records. Brunt argues that commanders did not take the time and trouble, especially following a defeat, to count the corpses.[59] If a defeated legion had to withdraw swiftly from the field, there was not time to perform this enumeration. Polybius' account of Zama

---

54 Leeman (1963) 176–7. It is ten times the frequency found in Cicero.
55 Livy 22.56.1.
56 *FRHist* 25: casualty figures F23, F27, F33, F34, F35, F36, F38, F39, F40, F43, F45, F47, F48, F65, F66a, F66b; treasury figures F28, F37, F51, F54, F62, F63.
57 Pelikan Pittenger (2008) 114. See later for more on numbers in a triumphal context.
58 Pliny, *HN* 33.55. Jacobsthal (1943) 307, distrustful of Antias, argues that Livy's triumphal details come from *aerarium* records when not 'invented' by Valerius.
59 Brunt (1971) 694.

indicates how much carnage there could be following a battle;[60] it may not have been possible to identify all the bodies, and would have taken considerable time. On the other hand, the ancient historians provide evidence of commanders burying the dead.[61] That mentions of this activity are usually used to contextualise other events suggests that it was routine. The first-century AD tactician Onasander even went so far as to consider it the primary responsibility of a commander.[62] This does not imply that a body count took place, but it does suggest that it was possible, should opportunity and desire coincide.

A fragment of Cato's *Origines* provides insight here:

> *Cum saucius multifariam ibi factus esset, tamen volnus capiti nullum euenit, eumque inter mortuos defetigatum uolneribus atque quod sanguen eius defluxerat cognouere. Eum sustulere, isque conualuit.*
>
> (*FRHist* 5 F76=Gellius, *NA* 3.7.19)

> While he [Q. Caedicius, a military tribune] had been wounded in many places during the battle, but he received no head wound, and they recognised him among the dead, worn out from wounds and because his blood had flowed out. They lifted him up, and he recovered.

The passage refers to the aftermath of a battle in the First Punic War in which a military tribune, Q. Caedicius, led what was effectively a suicide mission of 400 men in order to save the legion from defeat. Caedicius was found during a search of the battlefield. The passage has received a lot of attention in modern scholarship, largely focused on the linguistic and narratological features of this early Latin quotation.[63] In particular, the story is remarkably similar to that of M. Calpurnius Flamma and seems to have formed a topos concerning self-sacrifice and *devotio* in Latin historiography.[64] The events themselves have received little treatment. Here the importance is that the act of examining the dead could become part of a topos. This suggests that searching among the dead was a normal activity.

However, Cato does not reveal the exact purpose of the search, most likely because he believed his readership were already familiar with the process. Was the intention to find survivors, or merely to deal with the dead? The choice to move Caedicius despite his apparent death may indicate a special care for him. If so, this suggests a particular concern for discovering

---

60 Polybius 15.14.1–3.
61 Livy 23.36.4, 23.46.5, 27.2.9, 37.44.3; Plutarch, *Aem.* 22.5; Appian, *Syr.* 6.36.
62 Onasander, *Strategos* 36.
63 Basanoff (1950) 260–1, (1951) 281–4; Goldberg (1986) 174–5; Calboli (1996) 18–22; *FRHist III* 121–4.
64 [Livy], *Periochae* 17; Livy 22.60.11; Fronto, *Ep.* 1.5, 4.5; Pliny, *HN* 22.11; Florus 1.18.3; Basanoff (1950) 260, (1951) 281.

the fate of more senior individuals, as seems to be reflected by the inclusion of senior names in casualty reports (see later). On the other hand, the burial or cremation of the dead was a duty to be undertaken by the survivors if time and tactics permitted.[65] To achieve this, all the dead needed to be moved, and they could be identified (if possible) at this point. In the case of Caedicius, the lack of a head wound allowed identification, and luckily coincided with his regaining consciousness.

Literarily, Cato places equal emphasis on Caedicius' location and wounds to create a plausible narrative while at the same time providing an exemplum of Roman virtue resulting in a pseudo-resurrection. For this discussion, Cato's narrative demonstrates that as much care as possible was taken to examine the bodies of the dead and find survivors. It was through this process that a Roman general was able to generate at least a number of casualties and, wounds allowing, identify them. Such action was not always possible, but the case of Caedicius coupled with other passing mentions of the practice nonetheless demonstrates that this effort was made when possible.

However, the ability to gain accurate casualty figures does not guarantee that generals transmitted them as assiduously. A plebiscite of 62 sought to ensure the reliability of casualty figures in dispatches.[66] The law is later than the period under discussion here, but it suggests that reporting casualties had been usual practice for some time and was subject to adjustment by generals, presumably to make their achievements seem more impressive. Orosius also makes the complaint that writers had a tendency to inflate enemy losses and deflate Roman losses.[67] In mitigation, this was in part an argument to support his aim of debunking the idea of a 'golden past', and Orosius was concerned more with histories than dispatches, but it nonetheless highlights some ancient scepticism over the reporting of casualty figures.

While demonstrating that casualty figures were a regular part of dispatches, this once again raises the question of their reliability. That Valerius Antias obtained the figures from records is no guarantee of accuracy if commanders were deliberately sending erroneous reports to the senate. On the other hand, the law's context should not be overlooked. In the previous year Pompey had defeated Mithridates VI;[68] the First Triumvirate was only two years away.[69] With the power plays of this decade's influential men, it is plausible that the tribunes were reacting to these immediate issues rather than a longer-term culture of falsification. This does not mean that earlier generals did not attempt the same type of self-aggrandisement, but it does

---

65 E.g., Livy 23.36.4, 23.46.5, 27.2.9, 37.44.3; Plutarch, *Aem.* 22.5; Appian, *Syr.* 6.36.
66 Valerius Maximus 2.8.1.
67 Orosius 4.20.7–9.
68 Appian, *Mith.* 111–2, *BCiv.* 2.9.
69 Appian, *BCiv.* 2.9.

suggest that it had not become a serious problem until the 60s. Thus, the law helps to demonstrate that accurate casualty figures were a regular feature in Mid-Republican dispatches.

It is possible to largely circumvent the problem raised by the law of 62, however. Rosenstein points out, to counter Brunt, that a general did not need to count the bodies of the dead to ascertain casualty numbers.[70] The number of living could be subtracted from the legion's original strength to give the number of deceased. Brunt highlights that this does not take into account any prisoners or deserters.[71] This is true, but it nonetheless demonstrates that commanders could calculate their operational strength reasonably quickly after a battle without needing to see the dead. For operational purposes, it was the number of those remaining in the legion that was important, not a breakdown of the dead, imprisoned and deserted.

Valerius Maximus' passage on the law of 62 also provides a possible insight into why casualty figures may have been falsified. Although relating specifically to enemy deaths, it is worth noting. Valerius states that in order to triumph, at least 5000 enemy must have been killed in one battle.[72] Beard and Brennan have convincingly argued that these 'rules' were flexible and based on precedent.[73] Nonetheless, generals may have wished to play safe and inflate their figures if casualties were below this level. Deflation of Roman casualties, as discussed by Orosius, could have been fuelled by the same concerns. However, inflation of enemy casualties in letters to the senate had a much smaller strategic impact on tactical operations than deflation of Roman casualties. As will be seen (III:iv), reinforcements were sent to the legions on the basis of casualty reports in dispatches. Deliberately understating the number of Roman dead would in the long term leave the legions below strength and thus disadvantage them. This does not mean that underreporting did not occur, especially on shorter campaigns, but it indicates that generals had reason not to. Attitudes towards reporting enemy and friendly dead were unlikely to have been the same, as they were not used for the same ends.

Another suggestion can be made relating to accuracy and the triumph. In 211 Marcellus was denied a triumph on the grounds that his legions had not returned to Rome with him.[74] Develin has argued that the words attributed to Marcellus by Livy suggest that he was aware the army's absence

---

70 Rosenstein (2004) 109.
71 Brunt (1971) 694.
72 Valerius Maximus 2.8.1 – *lege cautum est ne triumpharet nisi qui quinque milia hostium una acie cecidisset.*
73 Beard (2007) 209–11; Brennan (1996) 318; see esp. Livy 40.38.8–9. Contra Mommsen (1893) 144–55; Weinstock (1971) 60; Develin (1978) 435. See Vervaet (2014) 68–130 for the most recent treatment with full bibliography.
74 Livy 26.21.2.

would be an obstacle;[75] the return of the legions was a clear sign of a campaign's completion. However, the legions' presence also allowed the senate to confirm the commander's dispatches. Even if the legion was extremely loyal to the commander and willing to support his version of events, its size would reveal the true casualty figures, as would the records of individuals kept with the legion (see III:iii–v). Only with the full return of a legion, accompanied by its records, could the truth be judged and a triumph (perhaps) awarded. If this was a factor, it points to accuracy in dispatches; a commander hoping for a triumph would eventually be caught in a lie.

The nature of some casualty reports adds to the impression of a detailed legion list held within the legion (see I:v). The casualty reports sometimes included the names and ranks of the more senior deceased.[76] The lists read very much like reports of significant losses, suggesting that they may have been lifted more or less intact from a general's report.[77] These men's names, in particular military tribunes and quaestors, were more accessible to the commander than those of ordinary soldiers because they were fewer in number and personally known to legion's commander. More often notices do not include names, but they do occasionally give the impression that Livy's source may have contained them.[78] Nonetheless, the need to keep detailed records for pay suggests that a legion list was kept and updated with individual information (III:iii). Any count after a battle could include not only the number of survivors but their names. Those found to be missing could be identified from the legion list, and information sent to the senate. The inclusion of the higher ranks' names does not indicate that commanders regularly sent lists detailing all the dead, but such detail does imply that at the least a number was sent, along with details of the more prominent. In terms of military administration within the legion, the Livian casualty figures reveal that commanders had access to both the number and names of all those under their command who remained alive and uncaptured.

The Battle of Cannae provides a good test case for the operation of casualty reports. The sources provide divergent evidence for both the number of combatants and the number killed, with Livy and Polybius both at odds and

---

75 Develin (1978) 432.
76 Polybius 8.35; Livy 25.34.11, 25.36.13–4, 27.1.12–3, 27.27.7–9, 33.22.7–8, 33.36.4–5, 34.47.2, 35.5.14.
77 E.g., Livy 35.5.14 – 'more than five thousand soldiers, Roman and allied, were lost, twenty three centurions, four prefects of the allies and Marcus Genucius and Quintus and Marcus Marcius military tribunes of the second legion' *supra quinque milia militum, ipsorum aut sociorum, amissa, centuriones tres et viginti, praefecti socium quattour et M. Genucius et Q. et M. Marcii tribuni militum secundae legionis.*
78 Polybius 2.11, 10.32; Livy 21.59.8–10, 23.11.9 (a Carthaginian report), 27.1.12–3, 30.6.8–9, 30.18.14–5, 33.25.9, 34.47.2, 35.5.14, 38.24.8, 39.31.14–6; [Livy], *Periochae* 22.10; Plutarch, *Marc.* 24.3, 29.5–9; Appian, *Sam.* 4.6, *Han.* 4.25, *Pun.* 15.102.

internally inconsistent.[79] There was confusion on the subject even 50 years after Cannae. The numbers fielded and killed are not relevant here, and so will not be discussed.[80] The focus must fall on the free (i.e., not captured) survivors. Internal legion records were most probably destroyed or lost as a consequence of Hannibal's victory; only the number of survivors as enumerated after the battle is likely to reflect a senatorial record. Livy set this at 32,500, which, as Toynbee points out, fits with the formation of three legions from the survivors.[81] The total of captured and killed derives from a count not of the dead but of survivors. The correlation between the two numbers indicates that the senate was concerned with the operational force available, not with the large number of prisoners or dead. This accords with the conclusions concerning Varro's letter to the senate following the battle (II:ii) and with the more general conclusions about intralegion administration at issue here. The aftermath of Cannae was unusual because the senate (not the commander) compiled a list of the whole standing force, but the same principles appear to apply.[82] In an emergency on a macro scale, the senate fell back on the administrative form well established within the legions.

Through examination of the casualty figures it is possible to conclude that Republican Roman legions operated with a relatively high degree of bureaucracy. Although the extant casualty figures may not in all cases be entirely accurate – whether due to scribal error, artificial emendation or rounding by ancient writers – their presence in the histories reveals that Roman commanders were capable of and required to keep a detailed record of their men. Even if generals slightly misrepresented their achievements before the senate through casualty figures, particularly in the first century, they themselves understood the true nature of their forces. Thus, the legion necessarily encompassed a great deal of administration to keep it running efficiently. The actions following Cannae indicate that this administration was in place by the end of the third century, and may well originate substantially earlier.

iii   Calculating *stipendium*

The surviving evidence of the Middle Republic reveals more detailed record keeping than just the numbers of the living and information about senior

---

79 Polybius 3.117, 6.58; Livy 22.49.13–8, 22.52.1–4, 22.59.5–6, 22.60.14, 23.11.9; Appian, *Han.* 4.25.

80 E.g., De Sanctis (1917) 131–5; Toynbee (1965b) 67–8; Brunt (1971) 696; Lazenby (1978) 76–85; Sabin (1996) 67; Daly (2002) 26–9, 45, 202.

81 Toynbee (1965b) 67 using Livy 22.49.15–8. Polybius gives about 3500, 3.117.

82 It could be argued that the senate was innovating by compiling a list of all survivors in order to gain an accurate picture of their forces, rather than utilising a system already in place. However, for the senate to compile this list they required the initiative of the remaining senior officers in the field. This suggests that Scipio and Varro were not doing anything unusual in reviewing the survivors.

officers. Knowing the number of men in a legion was not just necessary for tactical or logistic reasons. More detailed information can be obtained by examining how the *stipendium* (military pay) of soldiers was calculated. The man responsible for this was the quaestor. The quaestor who accompanied each consul with his army was responsible for the army's finances, in a role roughly equivalent to a modern quartermaster.[83] It is not necessary here to discuss the origin and development of the quaestor's role; suffice it to say that it had been since its inception early in the Republic an administrative position linked to the use of state finances both in Rome and on campaign.[84] Rosenstein argues that a list of recipients was necessary for quaestors to perform their duties fully.[85] This section will examine the quaestor's role as paymaster in order to demonstrate that not just a list but a separate record was kept for each individual on campaign.

It is not at issue here when military pay was introduced and how (or if) it rose (or fell) before the changes of Julius Caesar.[86] Here it is not how much legionaries were paid that is significant but how this pay was calculated and recorded to ensure that each man received his due.[87] Thus the nature of the compensation given to Mid-Republican soldiers must be examined in detail. The best way to do this is through an investigation of the deductions made from military pay. Military pay was originally intended to compensate citizens against the cost of going on campaign.[88] Deductions were made against this for state-provided items; the individual citizen did not pay for them and thus did not require compensation for them. Once again, Polybius provides the best evidence:

Τοῖς δὲ Ῥωμαίοις τοῦ τε σίτου καὶ τῆς ἐσθῆτος, κἄν τινος ὅπλου προσδεηθῶσι, πάντων τούτων ὁ ταμίας τὴν τεταγμένην τιμὴν ἐκ τῶν ὀψωνίων ὑπολογίζεται.

(Polybius 6.39.15)

But for the Romans the quaestor takes account of the arranged price of both food and clothes from the salary, and any additional arms if they are required.

Nicolet argues that this passage demonstrates that military pay was remarkably fixed; food and arms would have a relatively stable price, and

---

83 Polybius 6.39.15, 10.19.1; Sallust, *Iug.* 29.4. As Erdkamp highlights, also Livy refers to a Carthaginian officer responsible for grain supply as quaestor, reflecting Roman practice, Erdkamp (1998) 103; Livy 25.13.10.

84 Cf. Tacitus, *Ann.* 11.22.4–6 (*ut rem militarem comitarentur*); Mommsen (1894) 223–9; Latte (1936) 24–33; Harris (1976) 92–106. See IV for the quaestor's larger financial role.

85 Rosenstein (2004) 109.

86 Cf., e.g., Brunt (1950) 50–71; Watson (1958) 113–20; Boren (1983) 427–60.

87 See IV:iv for further discussion of pay administration.

88 Livy 4.59.11; Brunt (1950) 50; Marchetti (1975) 246; Boren (1983) 430.

thus as the needs of soldiers were steady, the deductions would be easy to estimate.[89] However, this interpretation is at best overly simple and at worst completely false. While Polybius gives the items for which costs could be deducted, he does not indicate the relative costs of these items or whether there was a fixed price for each. Watson suggests that the reason Polybius does not give a total figure of the deductions is because the same deductions were not made for each man, particularly with regard to clothing and arms.[90] It is necessary to examine each item – food, clothing and arms – separately in order to build a larger picture of calculating the *stipendium*.

Food

As has already been discussed (III:i), food is a good place to start. However, before continuing, a problem of translation must be highlighted. Polybius uses the term σῖτος. Liddell and Scott list the definitions as grain, food made from grain and food in a broader sense.[91] It is not entirely clear how the word should be translated on this occasion. It is probable that the original sense was similar to that of the Latin *frumentum* (grain), but as will be seen, it need not be limited to it. It may well be that the ambiguity of meaning aids understanding in what legionaries were to have deducted from their pay. Grain provided the main foodstuff to be accounted for, but it was not the only one. As will be seen, Polybius could and did refer specifically to wheat and barley if necessary, supporting the hypothesis that σῖτος here is not limited to grain. The double meaning of σῖτος as both grain and food can be understood from Polybius' description.

On the face of it, food cost should have been the easiest of the three elements of deduction to calculate. As Polybius implies, the deduction for food was made for each individual at a set price. Each infantry man was provided with four *modii* of grain per month and would have paid this cost accordingly. However, the passage does not necessarily mean that this fixed price remained the same year to year. Rome may have received this grain as a tithe from Sicily or Sardinia (thus free of charge), but the contracts for its transport still had to be let.[92] Depending on the distance to be covered and the competition amongst contractors in any given year, the cost of this transportation likely varied. As military pay was originally intended as financial compensation, it follows that the state made these

---

89 Nicolet (2000) 81–2; Taylor (2017) 149.
90 Watson (1958) 118.
91 LSJ s.v. σῖτος.
92 Erdkamp (1998) 86 with Cicero, *Verr.* 2.3.163; contra Rickman (1980) 105–7. With four *modii* a month per man, a tithe of three million *modii* would support 13.89 legions of 4500 men or 10.45 legions of 6000 men. The unusual contract situation of 215 occurred precisely because the state was unable to pay up front, Livy 23.48.10–49.3.

deductions at cost price, effectively reimbursing itself. Polybius is not specific enough to provide evidence to support this, but the use of 'the arranged price' rather than just 'the price' suggests that the price did vary, which again points to cost price deductions as the norm.[93] It therefore seems that while the value of grain deductions may have been fixed in a single year, it was not over the longer term.

The variable price points to other factors to be taken into account. While soldiers probably covered the grain's shipment cost, it does not follow that they also received deductions for food foraged on campaign. Fruit, vegetables and meat appropriated from the local area had no cost for either the individual citizen or the state. The quaestor may have organised the distribution, but it is doubtful whether a deduction would have been made against the accounts of legionaries. The other side of this conclusion is that if produce was purchased locally, this was a cost to the state. On the hypothesis proposed here, the men would also have this cost deducted from their pay. Such an interpretation is not beyond the scope of Polybius' description of deductions. When rations are discussed elsewhere, wheat is specifically referred to, suggesting that if Polybius had meant wheat rather than food in this instance, he would have used the term.[94] Arguably, this is pushing the semantic argument too far, but as a military man (see I:i), Polybius would have been familiar with the military diet. His statement emphasises the 'arranged' price because the nature of food supply and its cost were dictated by the unique circumstances of each campaign. Deductions for food were not limited to grain in all circumstances. The quaestor was required to deduct from pay the price appropriate to the army's circumstances, not a set amount.

It is also worth considering whether grain acquired on campaign resulted in a deduction. This is a more complicated issue. As already discussed, legions usually employed several methods of supply over a campaign to suit immediate tactical requirements. (Cato's entirely forage-based Spanish campaign cannot be considered an ordinary example.) Each man received his four *modii* a month, but it did not always come from the same source. It is unlikely that foraged grain and resupply grain were kept separate. The aim of this monograph is to demonstrate that the Mid-Republican army was organised using a relatively complex bureaucracy, but it is probably going too far to suggest that when mixed grain sources were used, the cost to the individual was diluted appropriately. Moreover, it is unlikely that foraged grain constituted the major source in the majority of campaigns. The need to keep the army moving from region to region and the

93 Cf. Marchetti (1975) 247.
94 Polybius 6.39.13 – πυρός.

unpredictability of forage made magazine supply a much more reliable and necessary option.[95]

On balance, it seems likely that in each year there was a set cost for the deduction of wheat based largely on a calculation of transportation expenditure. This was most likely calculated by the magistrate who let the contracts and passed on to the quaestors who accompanied the legions. In addition, any state expenditure (centrally or locally) on salt, oil, wine, vinegar/sour wine, meat, fruit and vegetables would be deducted from the men in the legions to which they applied. Thus, even in the same year, each army would have different food costs based on its circumstances. The quaestor's deductions would be unique to each army, and necessarily calculated in the field. Polybius was unable to give a set figure for food deductions because there was not one.

Clothing

Polybius' second deduction is for clothing. While this is rarely discussed in modern scholarship, there is no reason to question his accuracy. When Livy records costs or materials paid to Rome by its defeated foes (or as the price of peace negotiations), tunics for the soldiers appear regularly alongside rations and covering *stipendia*.[96] These are late-fourth-century cases, but there is no need to believe that providing clothing fell out of practice. In 215 the Scipio brothers in Spain reported to the senate that they required new clothes for their men,[97] and in 211 there was a passing mention to reviewing a request for clothing from L. Marcius.[98] In 169 the consul Q. Marcius wrote to the senate requesting that 6000 togas and 30,000 tunics be sent to him in Macedonia.[99] Finally, in 123, C. Gracchus' *lex militaris* made the supply of clothing free for the legions,[100] implying that the cost of clothing had previously been deducted from pay. Together with Polybius' own knowledge of Roman military practice, these examples demonstrate that the provision of military clothing was the norm throughout the Middle Republic.

These examples reveal further information about the provision of clothing. In all these cases, the army had been abroad (or away from Rome) for at least a year, suggesting that replacements were needed. It seems likely that the men originally left in their own clothes; there is no evidence for state provision of

---

95 Cf. Erdkamp (1998) 50.
96 Livy 8.36.11, 9.41.7, 9.43.7, 9.43.21.
97 Livy 23.28.4 – *vestimentaque*.
98 Livy 26.2.4.
99 Livy 44.16.1–4.
100 Plutarch, C. *Gracch* 5.1. Gabba (1976) 7 argues that this was a consequence of a lowered property qualification. It is more likely that it was part of measures to win the people's support. In effect, the measure increased military pay, as deductions were reduced.

clothing at the outset of a campaign, nor is it probable for an army without a uniform. With the onset of the Hannibalic War, long campaigns of several years had become the norm. Wear and tear on clothing over such a period, particularly but not only from damage in battle, suggests that replacements were a necessary and regular feature of warfare.

The examples from 215, 211 and 169 indicate that it was unusual for the state to have to send clothes out to the provinces. In his letter, Q. Marcius makes the point of saying that grain had been acquired locally but clothing and cavalry horses needed to be provided by Rome. This may imply that in the usual course of things, clothing could be obtained in the field. Indeed, the Scipio brothers were able to source their requirements locally. This may well be why Livy found these examples, and particularly those of the Marcii, noteworthy.

Q. Marcius' letter provides further insight. The size of the order – 6000 togas and 30,000 tunics – suggests that he was attempting to clothe his entire force, including the allied contribution and the unknown number of other men involved in the successful running of the army. Marcius had in Macedon two legions of 6000 foot and 300 horse, along with an equal number of allies.[101] This gives a total of 25,200 men; 30,000 tunics probably allowed for the support staff. This indicates that Marcius knew the approximate size of his force. Moreover, a deduction could easily be made from each man, as the cost would be the same across the board (except for those who also received a toga). However, Livy's interest in the letter suggests something more significant. As with the second point, it suggests that clothes were usually obtained on an ad hoc basis as required. It is implausible that every legionary's clothes fell to tatters at the same time. It may have been easier for the state to apply the same cost to everyone by organising or paying for purchase in bulk, but Livy's interest suggests that this was not usual. Rather, deductions would still be made, but marked separately for each man as he required a new tunic. As in the case of deductions for food, it appears that Polybius deliberately emphasised the 'arranged' price as unique to each situation; the need for new clothing was dependent on the length and nature of the campaign, and the fortunes of the individual legionary.

Arms

The case of arms is the third element of pay to be examined, and the most controversial. Polybius' description is often seen as evidence that in the mid-second century the state was regularly providing arms and armour for all its soldiers.[102] Gabba and Marchetti take a more staggered approach. Gabba

---

101 Livy 43.12.3–5.
102 Brunt (1971) 405; Marchetti (1975) 247; Gabba (1976) 10; Rich (1983) 287.

argues the passage demonstrates that in theory arms should be obtained by the individual, but in practice the state provided them.[103] Marchetti sees the passage as demonstrating that while in the Second Punic War men enrolled with their own arms, by the second century this had changed to state provision.[104] Along with Brunt, he sees this as a result of a lowered property qualification.[105] However, this is an attempt to have the cake and eat it too. Book 6 of Polybius' *Histories* refers to both Polybius' own time and 216 (see I:i). There had been some changes to Roman warfare during this period, but not in the principles and organisation on which it was based. Therefore, Polybius' property qualification applies to both points in time. Another explanation of Polybius' statement is required.

As Rich points out, the passage refers to a cost for replacement arms, not for the initial provision of arms. He bases his conclusion that arms were provided by the state in the 160s on Polybius' earlier statement that the first reassembly following the *dilectus* was made unarmed.[106] This passage has already been discussed (I:v), but it is worth reiterating the conclusions here. Book 6 applies equally to the late third and mid-second centuries. The first reassembly of the legion took place unarmed because it was the occasion on which the men were split into the lines of *velites*, *hastati*, *principes* and *triarii*. Until this took place, the men did not know what arms they would need and thus could not arm themselves appropriately. Indeed, it is not until chapter 26, five chapters later, that the legion is ordered to arm. Chapter 21 does not provide evidence of regular state arming, and even chapter 26 is at best ambiguous over where the arms will be sourced.

Returning to the quoted passage, it is necessary to examine the grammar in order to understand the nuance of Polybius' meaning. The comment on arms is in the subjunctive preceded by κἀν, indicating that unlike food and clothes, the cost of arms was taken only if necessary. The use of the subjunctive rather than the indicative, as seen with food and clothes, suggests that arms were a less frequent or expected expense. The implication is that men were expected to arm themselves. Polybius' choice of vocabulary further supports this conclusion. The use of προσδέομαι ('I require besides') rather than just δέομαι ('I require') emphasises this sense of condition. This again implies that any cost for arms would be for replacement arms. As the organisation of the battle line was in large part based on wealth, with equipment becoming more expensive in the more senior lines, this should not be surprising.[107] It does not preclude the state from providing equipment initially, but it does suggest that this was not standard practice.

103 Gabba (1976) 10.
104 Marchetti (1975) 247.
105 Brunt (1971) 405.
106 Rich (1983) 287 n.1; Polybius 6.21.6–7.
107 Polybius 6.21.7–23.9.

Rather, deductions for arms were made if the state had to provide replacements following loss or irreparable damage in battle. The commander was duty bound to turn over captured wealth to the quaestor.[108] As the quaestor commanded the baggage train,[109] it is likely that other spoils also ended up under his authority. Operating in unfamiliar territory and under oath to hand over looted spoils to the commander,[110] the only way for a legionary to replace equipment would be through the quaestor.

It is at this point that the importance of record keeping and its likely detail becomes clearer. When an individual needed to replace, for example, a shield, he could do so by visiting the quaestor. The value of this item would need to be marked down by the quaestor against the name of this soldier, as it seems was usually done with clothing. Aulus Gellius reveals that a single spear was the only equipment item which could be kept if found by a soldier.[111] This was probably the item which most often needed replacing, and may help further explain why Polybius considered the deductions for equipment infrequent. Nonetheless, when a legionary was paid, not only the universal deductions for food but also the individual deductions for clothing and replacement equipment were taken from the total he was to receive. For the system to work equitably, a record of each individual's account was necessary. Polybius' description of deductions from pay reveals that it was necessary for a detailed list of individual members of the legion to be kept on campaign in order for pay to be properly calculated.

Pay scales

The final variable to consider is the pay scale itself. Polybius breaks down *stipendium* into three categories: the lowest for an infantryman, twice that for a centurion and higher still for a cavalryman.[112] This suggests that, at the very least, a quaestor would require a list of men broken into these categories, or marked on a legion list, in order to ensure that individuals were paid correctly. However, Polybius' breakdown does not give an indication of what other ranks were paid. For example, was an *optio* paid the same as an ordinary legionary, and a military tribune the same as a centurion? Polybius does not use the term 'centurion' here, instead referring to ταξίαρχοι. This is a term which means 'unit commanders' and could mean centurions. However, Polybius explicitly uses κεντυρίωναι alongside ταξίαρχοι earlier in book 6, suggesting that while centurions were ταξίαρχοι,

---

108 Polybius 10.19.1; Sallust, *Iug.* 29.5–6.
109 Polybius 6.32.8 with Roth (1999) 258.
110 Gellius, *NA* 16.4.2.
111 Gellius, *NA* 16.4.2.
112 Polybius 6.39.12.

ταξίαρχοι were not necessarily centurions.[113] This is complicated by the use of καὶ, which could mean either 'and' or 'or'. Ταξίαρχοι is usually translated to mean 'centurions' throughout Polybius' description of Rome's military system, but this loses the nuance.[114] Rather, it seems that Polybius is attributing the medium rate of pay to unit commanders, and specifically not just to centurions. It cannot be stated with certainty exactly whom Polybius is envisaging being paid at this rate, but a reasonable assumption might be all those classed as 'officers', that is, anyone of the rank of *optio* and above.[115]

On the other hand, several modern scholars have attempted to recreate ranks both above and below centurion with their own discrete pay grades. Using evidence of the empire, Speidel posits four pay grades: rank and file, *centurio legionis*, *primus ordo* and *primus pilus*.[116] Likewise, Breeze suggests three grades below centurion: basic *milites*/technicians and specialists, junior staff officers and senior staff officers (including the *optio* and standard-bearer).[117] Although developed for the empire, both reconstructions are reminiscent of the ranks that appear to be missing from Polybius. Brunt believes that there were increasing rates of pay for ranks above centurion in the Republic, despite admitting that there is no evidence for this.[118] However, while an absence of evidence is not evidence of absence, it is safer to base conclusions on the existing evidence rather than appealing ideas. Thus, with the foregoing interpretation of Polybius' middle category there is no reason to believe that these six pay grades existed in the Middle Republic. Whether they developed during the Principate is outside the scope of this work.

In this way, Polybius' description covers all those in the legion with only three pay ratings. Besides fitting neatly into the Roman currency system,[119] this rating system would have simplified the work of the quaestor. Pay calculations had a simple basis to allow easier computation of the total. Moreover, military pay was designed as compensation for services rendered to the state.[120] There was no need for many graduated rates of pay in a system where wealth and experience designated an individual's place and

113 Polybius 6.24.4.
114 E.g., Paton, Walbank & Habicht (2011) 359.
115 Did cavalry officers therefore receive a fourth higher rate of pay? As cavalry pay was higher to reflect the cost of taking horses and attendants on campaign, and cavalry service itself was restricted to Rome's wealthiest, such differentiation may not have been required.
116 Speidel (1992) 100–2.
117 Breeze (1971) 134.
118 Brunt (1950) 68.
119 Different interpretations of Polybius' figures have been given, but the overriding tendency is to link his figures to whole numbers and coin values, e.g., Watson (1958) 114–6; Boren (1983) 439ff; Rathbone (1993) 152; Rathbone (2007) 159; Taylor (2017) 145–6.
120 Livy 4.59.11; Brunt (1950) 50; Marchetti (1975) 246; Boren (1983) 430.

pay was designed to cover basic campaign expenses. The heavy infantry may have had greater expenses, particularly in terms of armour, but they were allotted that place precisely because they were better able to bear that additional cost.[121] Polybius is describing a system of pay based on a mentality almost entirely alien to that on which Speidel and Breeze's professional army of the Principate based its pay grades.

However, while the complexities to be dealt with by the quaestor reinforce the notion of pay as fair compensation and not a reward, the system nonetheless indicates that Rome was beginning to move away from this strict view. The ταξίαρχος pay grade is not necessary if the concern is only with compensation. The ταξίαρχοι as individuals required the same compensation as ordinary soldiers; nonetheless, they were paid double. The obvious reason for this is that their seniority brought with it greater responsibilities and a greater risk of death. Polybius states that each maniple had two leaders, so that there would always be a commander in case one was killed or fled.[122] It is emphasised that this was to cover all possibilities, but it does raise some interesting points. Firstly, it highlights the importance of unit leaders once the battle was underway. They are painted as the pins holding the maniple together. Secondly, concern about their death demonstrates that this was a risk. As the centurions operated at the front of their unit, this risk was greater than for an ordinary soldier, who was unlikely to have always been positioned there.[123] Thus in essence it appears that the higher pay for ταξίαρχοι was a form of danger money. Polybius' pay scale reveals that in the Middle Republic the mentality behind military pay was beginning to move towards reward from compensation. That there was only one senior pay grade demonstrates that Rome was far from reaching this, but it was nonetheless starting the process that would be sped by the development of the professional army.

Quaestors in the field were faced with three basic pay levels from which deductions could be made. The variable deductions, particularly for clothing and arms, suggest that few men would ever be paid exactly the same amount. The fortunes of war, such as the loss or damage of equipment, coupled with the strategic, geographic and climatic influences, meant that while campaign expenses for the state may have remained relatively stable year to year, for individuals their 'take-home' pay would have varied more greatly. Detailed records were necessary to keep track of these differences. Even if there was a steady fixed price for items, which it has been argued here was not the case, the irregularity and nonuniversal need for new clothing and equipment would have required an individual record for each man in the legion. Nonetheless, the system was much less complicated

---

121 Polybius 6.21.7–23.9.
122 Polybius 6.24.7.
123 Implied by Polybius 6.24.7–9.

than if the quaestor had had to cope with six different pay rates, especially if promotions were made in the field. For example, if an *optio* was promoted to the rank of centurion, he would not receive an increase in pay for which the quaestor needed to account. The individual remained within the level of the ταξίαρχοι. It was only if an ordinary soldier was promoted into the ταξίαρχοι that a change needed to be noted. This is not to say that a promotion would not be recorded on the legion list, but simply that in the majority of cases it did not affect the work of the quaestor as paymaster.

iv   *Supplementa*

The ancient evidence reveals that more than just individual pay records were kept on campaign. It has been argued previouslythat recruiters had access to information about the length of previous service for each individual (I:v). This section examines the *supplementa* (reinforcements) sent to legions and the subsequent dismissal of term-served veterans in order to argue that this information was also carried with the legion and supplemented with information on service performance. Such an examination also reveals the limits of Rome's bureaucratic complexity.

The issues of *supplementa* and dismissal for *emerita stipendia* are often entwined in the sources. As Walsh points out, supplementary troops were enrolled at least in part to replace casualties and men ready for discharge.[124] It is worth examining these two groups – casualties and men eligible for discharge – separately at first. First, casualties. It has been highlighted that commanders had a strong grasp of their legions' size. This enabled them to send a report to the senate detailing the number of men required as a *supplementum* to cover men lost on campaign (III:ii). This could have been an explicit figure of the number of men required; however, there is no evidence of this type of request. Alternatively, the request could have been implicit in the casualty figures themselves. The senate knew the number lost and thus how many were required to restore the legion to full strength.

Second, men eligible for discharge. *Supplementa* also allowed the discharging of men who had served their term, indicating a more sophisticated form of record keeping within the legion.[125] It can be argued that in cases like second-century Spain (where it appears that the longest continuous service term was ideally six years; see I:iii) this was achieved simply by knowing in which year men had been sent to Spain, which in effect would require the discharge of entire cohorts or maniples rather than individuals. However, even this simpler system, based on units rather than individuals, would require some kind of record keeping in order to function

---

124 Walsh (1994) 128.
125 E.g., Livy 9.24.1, 39.38.10–1, 40.36.7, 43.12.5–6, 44.21.5–8.

efficiently and prevent mutinies over long service.[126] Moreover, while the legions themselves remained on campaign for this long, this does not necessarily reflect all individuals. The Spanish legions received *supplementa*.[127] Thus not all the men had served for the same period. Unless all the new soldiers were placed in a separate unit, it was not possible to discharge units wholesale to remove those with *emerita stipendia*. In reality, it appears that in Spain the legions were retired after six years with all members dismissed, avoiding this problem. Nonetheless, Spain appears to be unusual in this respect; certainly the piecemeal dismissal of veterans is much more prevalent in the sources.[128] This suggests that the individual service records carried with the legion included the number of years served, and were updated as years passed on campaign.

Thus the ability to both replace casualties and dismiss those who had served six years with *supplementa* provides strong evidence for relatively detailed record keeping on campaign. However, *supplementa* were for the most part enrolled in Rome on the senate's instruction and dispatched to the commander. This raises the question of how the numbers to be enrolled were decided upon. Whether this coordination was organised through the commander's dispatches or by parallel records is not entirely clear from the sources. Nonetheless, with the added complication of casualty numbers to contend with (particularly if commanders did not always send a full list of the names of the dead), it seems likely that this was achieved through letters. This supposition is supported by Livy's description of much *supplementa* recruitment, where commanders (or consuls acting for them) are instructed to recruit whatever number they deem suitable.[129] As already suggested, the recruiter could have consulted the copies of previous dispatches to calculate the required reinforcements. The senatorial decree on *supplementa* recruitment effectively devolved the decision to those who had, or could gather, a clearer idea of the requirements.

However, this system does not account for the dismissal of men with *emerita stipendia*. It appears that these men were dismissed only with any surplus remaining after the dead had been replaced.[130] This has several implications. The lack of specificity may indicate that the recruiters were aware that, despite efforts to be accurate, casualty figures could be wrong. The only casualty figures transmitted in the sources are for set-piece battles (see III:ii). It is unclear whether these figures account for those who died only in the battle itself or also of wounds afterwards. Further, there is no evidence that notifications of deaths from illness and disease were sent to

---

126 Cf. Messer (1920).
127 Livy 37.50.11, 34.56.8; see I:iv.
128 E.g., Livy 26.28.7; 32.8.3; 39.38.10.
129 Livy 25.3.4, 26.1.12, 27.8.11, 27.22.6, 33.43.6.
130 Livy 39.38.10–1, 40.36.7, 43.12.5–6, 44.21.5–8 (replacing invalids).

the senate (although the legion itself needed a record). The lack of evidence does not demonstrate that this did not occur; as a normal part of ancient life and armies, it was probably not of interest to ancient writers. Nonetheless, it suggests that even if the casualty figures from battles were known, other losses were likely not recorded. The magistrate responsible was therefore required to use his own judgment to come to a total that would cover the legion's losses. This may well explain why *supplementa* always occur as a round number.[131] Thus, while more complex than often imagined, the distances and uncertainties involved did impose a limit on the functioning of military administration. Perfect parallel record keeping between the legions in the field and the records in Rome was not possible.

The secondary place of veteran dismissal is further emphasised by over-generous reinforcements. On one occasion, not only the *emerita stipendia* but those who had given good service, *forti opera*, had to be discharged to bring the legion down to size.[132] In 169 and 168, the legions were left oversize once the veterans had been dismissed.[133] Together, these instances suggest that the *supplementa* were not raised with the knowledge of (or with great concern for) the exact number of *emerita stipendia* each year. It appears that the number dismissed varied depending on how accurately the recruiting magistrate estimated the legion's total losses. However, significantly here, these instances indicate that legion records allowed commanders to identify those who had served their term. Moreover, the note on discharge for brave service also suggests that some kind of conduct record was kept. Again, this indicates that legions kept more than just a record of units.

Why did Rome not calculate the number with *emerita stipendia* more accurately? It is argued here that the census contained a declaration of military service and that the legion list composed at the *dilectus* maintained this information. A copy of this legion list was kept in Rome (I:v). Consequently, the magistrate in charge of recruiting a *supplementum* had access to the service of each man in the legion. Even if the list in Rome was not updated with extra years of service once the legion departed, it would be a simple matter to calculate the current term. Why this apparently did not happen is unclear, but a hypothesis may be proposed. The method of estimating casualties appears to have always led to an overestimate, probably quite deliberately. Thus it was expected that there would be an excess with which men with *emerita stipendia* could be dismissed. It also allowed for any additional casualties since the previous dispatch, as well as deaths from disease. This indicates that keeping the legion at strength was more

---

131 E.g., Polybius 18.20; Livy 32.8.2 (3000 foot, 300 horse), 32.28.10 (6000 foot, 300 horse), 35.20.4 (4000 foot, 150 horse).

132 Livy 40.36.11. Whether or not this was a collective reward makes little difference here. Either way it indicates some kind of service record.

133 Livy 43.12.3–4, 44.21.5–8.

important than maintaining a service limit of six years. Polybius' absolute maximum service term of 20 years supports this hypothesis.[134] Men could be kept in the field for longer than six years legally, allowing the senate to focus on replacing casualties. Moreover, this focus allowed for veteran dismissal, if more sporadically than was ideal for the men themselves. The overall lack of mutinies suggests that the system was able to function without the need to consult the legion lists every time reinforcements were required. Rome's bureaucracy was complex only to the extent that it was required.

There is further evidence to suggest that commanders had more detail about their men's service terms than just when their cohort or maniple had been formed. Occasionally, following large losses, forces were amalgamated. In the case of the Spanish legions following the deaths of Publius and Gnaeus Scipio, this was a wholesale amalgamation of the two armies into a single force.[135] Even with the garrisons recalled and more than a legion's worth of Roman and allied reinforcement with Gaius Nero, another 10,000 foot and 1000 horse were considered necessary to restore the force to strength.[136] In this case, knowing when cohorts had been formed was not enough, as joining the forces resulted in the formation of new units, with no guarantee that all the unit's men started their terms at the same time. As this took place during the Hannibalic War – when, as has been seen, the situation was not an ordinary one – the discharge time of these men is unclear (and in the case of those who accompanied Scipio Africanus to Africa, not for at least ten years).[137] Nonetheless, it demonstrates that in order to regularly dismiss those who had gained *emerita stipendia*, commanders needed more information than simply when each cohort was enlisted.

v *Lustra*

This provides an excellent point to discuss the chapter's final element: *lustra* (reviews). It is the contention here that *lustra* were regularly performed in the field by incoming commanders to take stock of their troops, enabling records to be updated and any mistakes corrected. Scipio Africanus' arrival in Spain to command the combined remnants should thus be an ideal example of this process.[138] Livy does not use the term *lustrum* in his

---

134 Polybius 6.19.3.

135 Livy 25.37.4. Another example is the creation of legions from the free survivors of Cannae. Scipio's ability to select the most experienced of these veterans from Sicily before heading to Africa further suggests some kind of service record held with the legions.

136 Livy 26.17–9.

137 Livy 31.49.5, cf. I:iii. All veterans were withdrawn from Spain in 205, but it does not follow that they were discharged, especially as Livy uses *deducere* rather than *dimittere*. Livy 29.1.21.

138 Livy 26.20.4; Appian, *Hisp.* 4.19. Even if Lucius Marcius had already created new records in his reorganisation of the two armies, a *lustrum* is still likely, Livy 25.37.4.

description, and the other source is Greek, but the problem is surmountable. The *lustrum* originated as a purification rite performed by the censors.[139] As the census began as a military review, the purification cleansed the army as much as the citizen body. Thus, a *lustrum* of the army in array in the field should not be surprising, especially as the centuries arrayed outside the *pomerium* on the *campus Martius* at the census' beginning were an army in the field. This is supported by examples of *lustra* in military contexts.[140] Thus, Scipio's review appears to fit into this context despite the lack of *lustrum*'s explicit use.

However, the existence of *lustra* in the field does not demonstrate an implicit administrative connection. *Lustra* are also found in contexts with no bureaucratic link. In an attempt to define Mars' sphere of influence, Rosivach argues that Mars was predominantly associated with lustration. The key ceremonies containing this rite occurred in March and October (the traditional beginning and end of the campaign season), both in connection with Mars.[141] Objects necessary for war were the recipients of purification. Rosivach connects these with the censorial *lustrum* on the *campus Martius*. His overall argument is not strong, but the evidence nonetheless appears to indicate a link between Mars, war and purification. Cato the Elder also mentions *lustra* in an agricultural context.[142] These examples suggest that *lustra* were religious purification rites with no necessary administrative function, and that the censorial *lustrum* was thus unusual in this feature.

Despite this, a bureaucratic element to military *lustra* need not be ruled out. Otto argues that *lustrum condere* ('to conduct the *lustrum*') – the formula always found in connection to the census – should be interpreted as the storage of the review documents (i.e., the census documents) in the *aerarium* (treasury).[143] (The storage site of documents is a separate issue and will be addressed in V:iii.) Ogilvie dismisses this interpretation as largely overlooking the religious element of the ceremony and 'playing fast and loose with the meaning of -(s)-tro-m'. However, the assumption that Otto's interpretation overlooks the religious element is unfounded. The use of *lustrum* in other contexts reveals that it had a strong religious connotation. *Lustrum condere* as Otto imagines it sanctified the separation of Roman and other that the census embodied, and was performed by storing the documents which were its physical embodiment. A broadening of *lustrum*'s meaning to purification as the language developed does not rule out such a specific original meaning in the context of the census.[144] As

139 Varro, *Ling* 6.86–7.
140 Livy 23.35.5, 38.12.2; Cicero, *Att.* 5.20.2; Caesar, *BAfr.* 75.
141 Rosivach (1983) 512.
142 Cato, *De Agr.* 141.
143 Otto (1916) 17–40.
144 Cf. Ogilvie (1961) 34–5; *OLD* s.v. *lustro*; cf. Hamp (1986) 362.

lustration on campaign continued the original sense of a military review, it is not implausible that these lustrations had an administrative function alongside the religious.

However, as Ogilvie points out, the form of *lustrum* used for the census, *lustrum condere*, is unique to it, suggesting that a *lustrum* outside this context did not carry the same connotations.[145] Nevertheless, this need not be an obstacle to interpreting military *lustra* as administrative in line with the census. The language of lustration may hold the answer. If Otto's interpretation of *lustrum condere* is correct, it would have been incorrect to use the phrase in a context on campaign away from Rome. The sense of procession and ritual deposit carried by *condere* was inappropriate in the field, where deposit in a sanctified location was impossible. *Lustra* on campaign could not be considered *lustrum condere* precisely because being on campaign prevented the deposit of documents. It does not rule out the presence of the documents themselves.

When narrating Scipio's actions upon his arrival in Spain, Livy does not refer to a review at all. Instead, he refers to Scipio 'having done all there was to be done'.[146] The implication is that Livy expected his reader to know what this was. It may have involved a *lustrum*, particularly if one was a regular feature of assuming command, but it cannot be stated with certainty. On the other hand, Appian states that Scipio 'παραλαβών τε τὴν ἐκεῖ στρατιάν καὶ οὓς ἦγεν ἐς ἐν συναγαγών ἐκάθηρε'.[147] This translates as 'taking the forces already there and joining them in one body with those he brought, [Scipio] performed a cleansing'. The key word is ἐκάθηρε. Richardson translates this as 'he performed a ritual cleansing'.[148] Again, this does not indicate administration, but the whole phrase must be taken into account. Appian explicitly links this cleansing with reorganising the legions. Thus, as with the census, the cleansing itself may have been a religious purification, but it was strongly associated with a practical administrative act.

This sense is supported by Plutarch when he describes the same phenomenon in 169 as 'τὸν εἰωθότα συντελέσας καθαρμὸν αὐτῆς καὶ τῶν πράξεων' ('after the usual purification and review of them').[149] Like Appian, he associates the cleansing with an administrative task. Unlike Appian, on the occasion described by Plutarch the legions were undergoing not a reorganisation but a review at the arrival of the new commander. Moreover, Plutarch presents the event as an ordinary procedure, suggesting that it was

145 Ogilvie (1965) 31.
146 Livy 26.20.4 – *Scipio omnibus quae adeunda agendaque erant.*
147 Appian, *Hisp* 4.19.
148 Richardson (2000) 29; cf. LSJ s.v. Καθαίρω.
149 Plutarch, *Aem.* 36.3. πρᾶξις is a strange choice of noun, but there is not a manuscript problem, and 'review' is the neatest translation here.

usually performed by many incoming generals. This joint emphasis on purification and review from Appian and Plutarch suggests that both actions were necessary for a new commander.

Appian and Plutarch do not use the same wording to describe the purification and review. This suggests that there was not a standard Greek phrase to describe the process, indicating instead that both authors were translating a Latin term for which Greek has no equivalent. The obvious Latin equivalent for their descriptions is *lustrum*. As both authors demonstrate, Greek has a term for purification, καθαίρω, but this alone was not enough to describe the process. This further indicates that *lustrum* in a military context has a larger meaning than just purification, encompassing an association with reviews and documentation. The sense of routine implied by all three authors suggests that the review was a normal feature of command. Thus it appears that an up-to-date record was of as much practical importance to commanders as the religious element of the *lustrum*.

To conclude, through the examination of records of the living it is possible to see bureaucracy working within the legions. The ability to dismiss men on the basis of *emerita stipendia* demonstrates that commanders in the field had some type of record of the service length of the men under their command. Coupled with the occasional reshuffle within legions to form new full units and the ability to dismiss men based on good conduct, this indicates that these records were not simply on a cohortal or manipular level but included, to some degree, service records of individual men. A new commander's military *lustrum* performed on arrival in his province was more reminiscent of the census' original form than what it had become by the Middle Republic. It provided an ideal opportunity to take full stock of the legion, dismissing surplus veterans and ensuring that records were as accurate as possible. Without these records, the commander in the field had less information about his forces' strength and composition, opening himself to possible mutiny if he was unable to deal with issues such as *emerita stipendia*. Military bureaucracy was just as necessary within the legion as at Rome.

vi   Tracking manpower on campaign

Overall, it can be seen that it was necessary for commanders to keep detailed records of the men under their command while in the field. This achieved several goals. Firstly, it was key for the successful operation of the legion. From a tactical perspective, knowing the true strength of the legion allowed a commander to commit his army more appropriately. It served no purpose to continue to assume that a legion contained 5000 men if it was known that losses had since been sustained. The frequent inclusion of casualty figures in dispatches to the senate reveals that this was an active concern of commanders. Further, understanding the legion's logistical requirements was directly linked to knowing the number to be supplied. This prevented over- or undersupply, allowing the legion to operate as efficiently as possible to reach the commander's strategic aims.

Secondly, detailed records of the legion's members allowed the state to keep track of its obligations, both financial and personal. Military pay still maintained its role as compensation for individual costs accrued while fulfilling an obligation to the state rather than reward for services rendered. Deductions from pay thus needed to be tracked to ensure that each citizen soldier received the correct compensation for his costs without the state being overcharged. Records of previous service carried over from the census list were also included, allowing the dismissal of those who had reached the end of their six-year term. In reality, this dismissal was somewhat ad hoc, but the process was nonetheless sufficient to keep the legion running smoothly.

It is now possible to understand the process of military administration on campaign and its interaction with the central military documents in Rome. In the field, commanders had a record of the men serving under them. As established in I, this record originated as the list of members drawn up at the *dilectus*, including previous service, and additional information on rank was added once the legion was organised. Using this list, the quaestor was able to calculate the appropriate pay for each individual, marking all the separate deductions to be made from each soldier. Commanders took care to keep the record of their numbers up-to-date, recording casualties in as much detail as time and injuries allowed. This information was transmitted to the senate in order to keep the legions up to strength. Further, the detail in the legion's records allowed not only casualties to be replaced but also those who had achieved their *emerita stipendia*. Undue service requirements could for the most part be avoided with this system, preventing long campaigns that kept men in the field for much longer than the ideal limit of six years. The complexity of the records and the organisation required to keep this bureaucracy and thus the army operating smoothly suggests that every effort was made to keep the records as up-to-date as possible. However, as with any bureaucracy, circumstances of campaign meant that there could be errors. Thus the *lustra* conducted by new generals provided the opportunity not only to ritually purify the army under the auspices of a new commander but also to take stock of the army and update the records in case of any omissions.

All this demonstrates that, within the legion and in Rome, detailed records were kept, and, within reason, every effort was made to keep them as accurate as possible. The previous chapter demonstrated that Rome, as might be expected, was the military administrative hub. The census declarations and census list served as the central authority for military records. Lists such as the *tabulae iuniorum* could be created from it, with exemptions and service terms noted. From these, at the *dilectus*, legion lists including the same details could be created. A copy of this list was left in Rome and another taken with the legion. These parallel documents enabled a degree of cooperation between the administrative authorities within the legions and at Rome. The legion lists allowed commanders (or their subordinates) to act as devolved satellite bureaucracies, with more exact information from being on the spot. Frequent letters and embassies from the legions to the senate meant that these satellites could communicate not only their tactical position but also administrative information. Roman military bureaucracy in the Middle Republic was a complicated and layered affair.

The inexact *supplementa* and the possibility of the dead being included in the census by their unknowing *paterfamilias* add to this picture of insufficiency. However, these problems were more a result of distance and travel time for messages than a major administrative failing. Rome endeavoured to keep its records updated, but it was only when a commander and legion were recalled and dismissed that the records could be fully matched up. The new commander's *lustrum* was an opportunity to update records in the field, but this information may not have been available to Rome if the previous commander had left before his successor arrived. As this is a technological limitation rather than a bureaucratic one, it should not take too much away from Rome's achievement in creating a complex and flexible military bureaucracy in order to keep track of manpower at home and in the field.

4 *Tributum* and *stipendium*

There is one final area of military administration still to be examined. Moving away from the records of soldiers on campaign, it is necessary to discuss the larger-scale processes which financed Rome's military endeavours. For the most part, this came in the shape of Rome's only form of direct taxation: *tributum*. *Tributum* was a proportional tax applied to individuals on the basis of their census rating. It was calculated by taking the costs of campaign, comparing this with the total census ratings, expressing the difference as a percentage and then taxing accordingly. Everyone paid the same percentage of their landed wealth.[1]

The tax was technically levied only as necessary, but by the Middle Republic it had effectively become a regular, yearly tax until its cessation in 167.[2] The proceeds were used to fund *stipendium* (military pay), which was by far Rome's largest war cost;[3] the money from *tributum* covered the majority of a campaign's costs. Therefore, understanding the mechanisms by which *tributum* was arranged, collected and then distributed as pay provides a crucial part of the picture of Mid-Republican military bureaucracy. As with other elements of this administration, the Middle Republic promoted development and introduced centralisation, but in many respects to a lesser degree than in other areas. This chapter examines the processes of *tributum*. After a note on terminology, it establishes exactly who paid *tributum* and at what rates. Following that, the collection of *tributum* is examined, demonstrating the important role of the *tribuni aerarii* in local tribal networks and in minimising the administrative burden of the state.

---

1 Varro, *Ling. Lat.* 5.181; Dion. Hal. *Ant. Rom.* 4.19.1–4 with Nicolet (1976) 11–21, 35–45. There is no basis for believing the proportion to be fixed, contra Frank (1932) 2; Frank (1933) 66; Marchetti (1977) 131. See Rosenstein (2016b) 90 for two possible reconstructions of payments.
2 Cicero, *De Off.* 2.76; Val. Max. 4.3.8; Pliny, *NH* 33.56; Plutarch, *Aem. Paul.* 38. Note that *tributum* was halted but not abolished.
3 See Frank (1933) 95–6, 145–6, Rosenstein (2016b) 81 and Taylor (2017) 145–7 for estimated cost breakdowns. The soldiers themselves never saw the full amount, due to the deductions from their pay; see III:iii.

Discussing the rare instances of *tributum* refunds sheds further light. Next, the payment of *stipendium* is examined to reveal the technically small but conceptually large changes required as Rome's wars grew in scale and scope. Finally, the investigation concludes by examining the only surviving set of quaestorial accounts, which record the *tributum/stipendium* payment to the quaestor. The discussion focuses on process; exact values are of less interest here than the mechanisms which enabled it.[4]

i Terminology

A brief note on terminology is required before continuing. Despite the modern convention to refer to the direct tax used to fund the *stipendium* as *tributum*, it is not the only meaning of the word in Latin. *Tributum* is also used: to refer to indemnity payments, such as those from Carthage; for contributions to purchase peace talks; and later for taxation of the provinces.[5] The meaning of the term as required here must be inferred from context, thus opening some scope for varied interpretation. Additionally, the term *stipendium*, used here to refer to military pay, also has a variety of meanings. It is often used interchangeably with *tributum* to refer either to the direct tax or to the other meanings of *tributum*, and occurs much more frequently than *tributum*; it also came to mean campaign years.[6] Again, this allows some variation in interpretation in each case. However, the close relationship of *tributum* and *stipendium* is also of help here. It demonstrates their interrelation in the mind of even those authors writing after *tributum* ceased to be levied. For example, in 209 the Latins complained of being exhausted by military and financial levies, and Lintott notes that down to the Social War the *tributum* levied on non-Romans in Italy remained 'chiefly, if not solely' related to providing pay for allies.[7] That *stipendium* can mean either military pay or the tax (or indemnity) that funded it indicates that the money was clearly earmarked for a certain purpose discrete from other state revenues.[8]

For clarity, it must also be noted that whilst the terms *aerarium* (treasury) and *tribuni aerarii* (tax tribunes) look almost identical, they do not carry the same meaning. *Tribuni aerarii* should not be translated as treasury tribunes. Both share an etymological route relating to *aera* (bronze), but each developed separately to the other.

---

4 Cf. Frank (1933), Rosenstein (2016b) and Taylor (2017) for discussions of the financial numbers and further reading.
5 *OLD* s.v. *tributum*.
6 *OLD* s.v. *stipendium*.
7 Livy 27.9.2 – *dilectibus, stipendiis se exhaustos esse*; Lintott (1993) 70.
8 For further discussion of the meanings of *stipendium* and *tributum* and derivative terms in the Republic and Principate, see France (2006); Soraci (2010); Ñaco del Hoyo (2019).

ii Taxpayers and tax rates

The definition of who was to pay the *tributum* appears to have been clearly laid out at its inception. It is generally accepted that *tributum* began alongside the creation of *stipendium* during the war with Veii.[9] Livy narrates that the tax was to be paid by those liable for military service but not serving in that year, in order to cover the costs of those who were.[10] The principle of either/or liability was confirmed in 401, when *seniores* serving as a garrison in Rome objected to *tributum* demands on the basis that they were themselves serving.[11] This appears very clear-cut: any citizen liable for military service was also liable for *tributum*. However, on closer inspection the military and financial liabilities of the *assidui* are not identical. For example, unlike with military service, there was no maximum number of years (theoretical or otherwise) for which *tributum* was to be borne. It is by examining these differences and their nuances that a more detailed picture of *tributum* and its limitations emerges.

Livy's description suggests that all *assidui*, both *iuniores* and *seniores*, had the same *tributum* liability. However, modern scholars are in agreement that only those *sui iuris* (under their own authority, i.e., a *paterfamilias*) had to pay *tributum*.[12] This may seem a small distinction, but it is a significant one. It means that adult sons still under the authority of their father were liable for military service but not for the tax which funded it. For example, a household with three adults liable for military service would only be liable for one *tributum* payment. As Rosenstein highlights, adult sons under their father's authority were technically *proletarii*, as they owned no property.[13] Whilst their father's status was 'borrowed' for military service, this was not true for financial liabilities. This is logical. The individual's tax rate was based on his census rating, which in turn was based on the property of the *paterfamilias*.[14] To tax the three-adult household by head would be to triple its tax burden. As will be seen throughout this chapter, the ancient evidence is not forthcoming about the details of Rome's only direct tax. However, that such a distinction between financial and military liability within the same household could be and was made is demonstrated by the creation of a list of taxpayers derived from the census (see II:v). Military *assidui* and financial *assidui* were not identical, and Rome's system was able to encompass this.

---

9 Ogilvie (1965) 622; Nicolet (1976) 18 (sometime in the fourth century); Crawford (1985) 21; Mohr Mersing (2007) 216, 229–35; Northwood (2008) 265n.30; Tan (forthcoming). Contra Brunt (1971) 64; Ñaco del Hoyo (2011) 382.

10 Livy 4.59.11–60.8.

11 Livy 5.10.4–5.

12 Brunt (1971) 15–16; Nicolet (1976) 29; Mohr Mersing (2007) 216; Rosenstein (2016b) 84; Tan (forthcoming).

13 Rosenstein (2016b) 84.

14 Varro, *Ling. Lat.* 5.181.

Significantly, this distinction between financial and military obligations also indicates that Rome's potential needs and potential provision were fundamentally unbalanced. The tax base to fund military endeavours would never be as large as the manpower base to fight them. This lends weight to the argument that the 338 enfranchisement was more about gaining tax-payers to spread the tax burden than gaining manpower to increase Rome's reach.[15] The creation of the *cives sine suffragio* category of citizenship in 338 allowed Rome to regularise payments from former enemies; simultaneously, it allowed more full citizens to be recruited into the army without increasing the tax burden on the other *assidui*.[16] On the other hand, I:iv has shown that the actual available manpower reserves could be much lower than headline census figures (or even extrapolated *iuniores* numbers) might suggest, and that the senate knew this. However, despite this, Rome always had the potential to struggle to meet the basic *stipendium* costs. If the enfranchisement of 338 was about tax, it may show a developing awareness, but the reality was that expanding the tax base also expanded manpower. The problem was put off, not solved. As with all of Rome's administrative systems, the *tributum* system's flaws only became fully evident and influential in periods of unanticipated pressure. Thus it was only during periods of unprecedented mobilisation, in the First and then more severely the Second Punic War, that the taxpayer/manpower balance was so unequal that the consequences manifested on a large scale.[17]

Livy's account of stress on the *assidui* in 215 supports this interpretation. Livy reports that following the battles at Trasimene and Cannae the numbers available to pay *tributum* were much reduced, placing a much greater burden on those remaining.[18] In other words, as the tax pool shrank, individual households found themselves liable for ever-increasing contributions in order to meet *stipendia* costs. At first glance, it appears that Livy is referring to the high number of casualties; both Trasimene and Cannae resulted in high loses for Rome.[19] However, it has already been demonstrated that as many as 50% of serving men had a living *paterfamilias*.[20] The deaths of these men should not have affected the tax base. Further, the deaths of some would have created a new *paterfamilias*, transferring the liability to the surviving son but not removing the household from the tax pool. Whilst it is impossible to quantify where the losses fell within these demographics, the overall effect on the tax base should not have been as severe as Livy suggests. Other events

---

15 Livy 8.14.1–4; Rosenstein (2016b) 95; Taylor (2017) 170n.82. Cf. Rosenstein (2012) 108; Tan (2020) 53.
16 Rosenstein (forthcoming).
17 During the First Punic War, Roman citizens twice legislated to ban publicly funded fleets. 253: Polybius 1.39.2; Zon. 8.14–16. 249: Polybius 1.55.2; Zon. 8.16. Cf. Tan (2015) 212.
18 Livy 23.48.4.
19 Trasimene: Polybius 3.84; Livy 22.7. Cannae: Polybius 3.117; Livy 22.49.
20 See II:ii.

help explain 215's tax difficulties. Despite recurrent problems with recruitment, Rome still raised new troops. This included 5000 foot and 400 horse to be sent to Sardinia, in addition to mobilising two legions enrolled in the previous year.[21] While this was not significant recruitment in the context of the Hannibalic War, the cumulative effect of four years of enlistment, reinforcements, no discharges and deaths was finally beginning to make itself felt as a reduced number of taxpayers. When this is combined with other factors, such as farming pressures and depleted household reserves, it is unsurprising that taxpayers began to struggle as the increasing burden was split between fewer and fewer.

As the payment of *tributum* was roughly equivalent to military service, this raises a question about the status of the *senes*, men over 60 who were not liable for military service of any type. On the basis of this equivalence, they should not have been liable for *tributum*. However, it is highly unlikely that this group, the vast majority (if not all) of whom were *patresfamilias*, was excluded from the tax base simply on the basis of age; age is a limiting factor for physical military service, but not for financial payments. Indeed, Brunt argues that *senes* had to be registered in the census precisely for the purpose of taxation.[22] The author is unaware of any direct evidence, but the balance of probability is strongly in favour of the payment of *tributum* by *senes*.

A final group of interest when considering *tributum* taxpayers is priests and augurs. Livy reports that in 196 the urban quaestors demanded money from the priests and augurs which they had not contributed during the Hannibalic War.[23] As Briscoe notes, it is unclear whether priests were exempt by custom or simply had not paid; that the quaestors could demand the arrears demonstrates that they were not exempted by law.[24] All that can be said here is that if the systems of collection were as argued later (IV:iii), the failure of the priests and augurs to pay *tributum* was as much a failure of collection as contribution. This suggests that their lack of contribution was sanctioned by other upper-class Romans, indicating a custom. In a period when special commissions were created to seek out unpaid tax, it is unlikely that priests and augurs were accidentally overlooked.[25]

However, the case of the priests and augurs does help to highlight the limits of Rome's tax system. Priests and augurs would have been of senatorial class and potentially amongst some of the richest men in Rome; in 196 they were apparently able to pay 18 years of arrears at once. The lack of their contribution during the war would have been significant, indicated

---

21 Discussions on recruitment: Livy 23.24.5; legion for Sardinia: Livy 23.34.13; mobilised legions: Livy 23.31.5.
22 Brunt (1971) 21.
23 Livy 33.42.4 – *quaestores ab auguribus pontificibusque quod stipendium per bellum non centuilissent petebant*
24 Briscoe (1973) 329.
25 216–10: Livy 23.12.6; 24.18.12; 26.36.8.

by the use of the cumulative amount to pay off state debt. This case illustrates a point made by Kay: despite the Roman state's liquidity problems during the Second Punic War, there remained huge pools of private wealth. In 210 Scipio was able to find 400 talents of private money to take to Spain, while in 205 individuals were rich enough to purchase swathes of the Campanian territory.[26] That *tributum* was levied proportionally by quota across the population prevented Rome from accessing this wealth. Unlike modern progressive taxation, where those with more pay a greater percentage, with *tributum* those at the top of the first class paid the same proportion of their property wealth as those at the bottom of the fifth class. There was no mechanism to access a greater proportion of their wealth without putting the same, intolerable, burden on the poorest taxpayers.

Moreover, as the failure to address this problem even during the Hannibalic War demonstrates, the Roman elite had little incentive to redress this balance. Rather, public funds relied on extraordinary contracts,[27] double taxation[28] and (more or less) voluntary contributions to make up the shortfall.[29] Several factors influenced this. Firstly, given the short-term nature of magistracies, any increase to the state's power to solve a short-term problem could well be to the benefit of a rival further down the line. Secondly, Roman elites had little desire to spend their private wealth for public gain unless they received recognition in return. The games and temples of victorious generals were paid for from their own pockets but also enhanced their reputations, as Livy's multiple notices demonstrate.[30] These acts were exchanges, not gifts.

Thirdly (and perhaps most importantly), while the introduction of varied tax rates could be viewed in line with the greater military obligations of richer and more politically ambitious citizens, it would have required greater bureaucracy. More detail would have been required than just the total to be collected and a list of taxpayers with their census class. The greater the differentiation, the greater the administration. The previous chapters have demonstrated that Rome was capable of developing more complex systems in times of strain. However, *tributum* provides a counter-example. The extraordinary measures of the Hannibalic War allowed military finance to cling on; the drastic measure of retariffing the coinage rendered changes to *tributum* ratings unnecessary (temporarily at least). In this unusual case, the bureaucratic legacy of a much smaller Rome was able to continue supporting its expansion unaltered.

---

26 Kay (2014) 17; Polybius 10.19.2; Livy 27.36.12, 28.46.4–6.
27 Livy 23.49.1–3.
28 E.g., Livy 23.31.1.
29 E.g., 24.18.15. Serving men also refused their pay with varying degrees of enthusiasm, Livy 26.35.2.
30 See Ziowlkowski (1992) on Livy's temple-building notices.

iii Collecting *tributum*

Having established who paid *tributum*, the next step in the process was to collect it. It is through the collection of *tributum* that much of the administrative complexity of military finance can be observed. This section will examine the role of the *tribuni aerarii* (tax tribunes) in order to demonstrate that Rome was able to operate an efficient direct taxation system with minimal input from central magistrates or institutions. Moreover, rare examples of this system operating in reverse, when the *tributum* was refunded, shed further light on both the financial side of military bureaucracy and broader Roman society.

Exactly who collected *tributum* and how this mechanism operated is extremely difficult to glean in the ancient sources. As the collection of *tributum* from citizens was halted in 167, none of the extant sources had contemporary experience of its operation.[31] Polybius was an adult, but not yet in Rome; even his keen interest in the detail of Rome's military systems seems to have stopped short of its financial backing. Lost writers such as Fabius Pictor did experience *tributum*, but it is questionable how much interest they had in describing their tax system, or how well any references would have been understood by their post-167 literary descendants.

There is a description of the *tributum* system provided by Dionysius of Halicarnassus.[32] The account is problematic for several reasons, not least due to his following the Roman tradition of anachronistically attributing direct taxation to the sixth century. More importantly here, Dionysius suggests that *tributum* is equivalent to the *eisphora* of fourth-century Athens. In brief, the *eisphora* required those liable for the tax to be split into 100 groups, each liable for 1% of the total tax burden. Within each group, the three wealthiest members paid the tax upfront and the recovered the cost from other members.[33] For Dionyisus, the 193 centuries of the centuriate assembly formed each of the groups for *tributum*. Nicolet argues that Dionysius' use of fourth-century Athenian terms that were not contemporary to him renders this a reliable description.[34] Other scholars are more cautious, arguing that it is not enough to simply equate *tributum* with *eisphora*.[35] Moreover, by following Dionysius, Nicolet runs up against the thorny problem of how the tribes and centuries operated together. This is a difficulty which must be overcome if Dionysius' description is to be accepted. It is better to focus on other evidence where possible.

Fortunately, Varro describes the collection of *tributum* when he discusses the term *tribus* (tribe):

---

31 Cicero, *De Off.* 2.76; Val. Max. 4.3.8; Pliny. *NH* 33.56; Plutarch, *Aem. Paul.* 38.
32 Dion. Hal. *Ant. Rom.* 4.19.1–4.
33 On *eisphora* see Christ (2007); Demosthenes 42.3–5, 42.15; Demosthenes 50.8–9.
34 Nicolet (1980) 161. Cornell (1995) 197 also accepts Dionysius' description.
35 Marchetti (1977) 108; Tan (forthcoming).

> *Tributum dictum a tribubus, quod ea pecunia, quae popula imperata*
> *erat, tributum a singulis pro portione census exigebatur. Ab hoc ea*
> *quae assignata erat altributum dictum; ab eo quoque quibus altributa*
> *erat pecuni, ut militia reddant, tribune aerarii dicti.*
>
> (Varro, Ling Lat. 5.181)

> *Tributum* was said to be from *tribus* [tribe], because that money, which
> was ordered from the people, was taken tribe by tribe from the individual
> in proportion of the census. From this, that which was assigned was called
> allotted [lit. attributed]; also from this, those to whom the money was
> attributed, that they might give it the soldiery, were called *tribuni aerarii*.

Varro's account provides one of the few mentions of *tribuni aerarii*, the
individuals responsible for collecting *tributum*. These men are almost invisible in the historical record, but their important status is not in doubt.
The *ordo tribuni aerarii* appear in Cicero's speeches, where he recounts
them being called to arms in 100.[36] The *ordo* appears to have been similar
in status to the *ordo publicani*, perhaps unsurprising if their tax-collecting
activity provided the eventual model for *publicani* and tax farming in the
provinces.[37] *Tribuni aerarii* are also visible in Plautus, where a soldier can
be seen approaching a rich man asking for his pay.[38] The term *tribunus*
aerarius is not used, but the scene seems to correspond with a soldier's legal
right to *pignoriscapio* (property seizure) to recover his pay.[39] Together,
these examples sketch the role and place of *tribuni aerarii* within Roman
society. They were rich men, as Plautus attests, trusted with the collection
(and distribution) of Rome's most important source of military finance.
They were also prominent within society; men knew from whom they
should collect their pay. Their social status was such that their identity as
tribuni aerarii long outlasted their practical function.

Using this evidence, modern scholars have recently described the collection of *tributum* in the fourth and early third centuries. Upon the end of a
campaign and the return of men to Rome, the campaign's costs would be
calculated and the tribes informed of their share. This entire cost was then
paid by the *tribuni aerarii* to the men individually as they came to collect
their entitlement. *Tributum* was collected from individual citizens within
the tribe not as a single payment but in a series of ongoing transactions
between richer and poorer farmers throughout the year.[40] The men who

---

36 Cicero, *Pro Plancio* 21.
37 Cf. VanDerPuy (2020).
38 Plautus, *Poenulus* 1285–6; *Aurularia* 525ff.
39 Gaius, *Inst.* 4.26; Aul. Gell. *NA* 6.10.2–3; Hill (1946) 61; Tan (forthcoming). Interestingly,
   publicani had the same right against defaulters.
40 Tan (2017) xxiv; Tan (forthcoming).

became *tribuni aerarii* were already powerful local landowners embedded in patronage networks before *tributum*. Their position meant they had the access required to collect the levy as a local intermediary.[41] This added responsibility required only a slight flexing of existing muscles. Thus *tributum* was integrated into the normal fabric and exchange mechanisms of Roman society.

This reconstruction has many merits. It helps to explain the crucial but largely invisible role that the tribes played in Roman society, as well as revealing the operations of powerful, but nonmagisterial, elites. It also means that any objections to *tributum* on the grounds of lack of coinage are overcome.[42] Only the final payment to the soldier needed to be in coin; many of the other transactions would have taken place in kind as part of the regular exchange mechanisms. Further, from an administrative perspective, the central state has very little involvement beyond informing each tribe of its dues; as Tan puts it, 'the state left the locals to figure it out'.[43] This accords entirely with the picture emerging in this monograph of a centralised state administration kept as lean as possible.

Despite this, there are elements of the reconstruction which require further comment. Nicolet suggests that campaign costs were estimated at the beginning of the year, but Tan argues that this was not necessary.[44] *Tribuni aerarii* would gradually collect the tax over the year and pay the exact amount at the end. This is not problematic for the shorter campaigns of the fourth century, but its efficacy must be questioned as campaigns grew in scale and scope through the Middle Republic. As already mentioned, *tributum* covered the majority of Rome's military costs. Whilst the men themselves may not have been paid until the end of the year, the costs they accrued (food, clothing, arms, transport, etc.) were required before then. In times of great military and financial pressure, Rome did not have residual funds to cover these up-front costs. In 215, three groups of contractors agreed to supply the Spanish legions only on the conditions that they be exempted from military service, the state underwrite their ships and they be first in line for repayment. Livy is explicit that the state was now funded by private money.[45] However, this in itself demonstrates how unusual it was

41 VanDerPuy (2020).
42 Brunt (1971) 64; Ñaco del Hoyo (2011) 382.
43 Tan (forthcoming).
44 Nicolet (1976) 20; Tan (forthcoming).
45 Livy 23.49.1–3 – *privataque pecunia res publica administrata est*. Some scholars consider the appearance of private contractors (*publicani*) in 215 anachronistic: Erdkamp (1995) 171; Erdkamp (1998) 117–19; Ñaco del Hoyo (pers. comm.); contra Badian (1976) 16–21; Richardson (1986) 39. However, Prag has highlighted that private contractors were responsible for shipbuilding as early as 267: Lydus *de mag.* 1.27 with Prag (2014) 200. On balance, it is likely the term *publicani* which is anachronistic in 215, not the presence of private contractors who would in the following century form the *ordo publicani*.

for the state not to have financial reserves. Moreover, the volume of other evidence suggests that *tributum* was regularly levied at the end of the year. There are no notices of *stipendium* funds being taken as spoils when Roman camps were captured, indicating that such large sums of money were not regularly carried with the legions; such a loss should have attracted comment. Enemy indemnities to cover campaign costs were by their nature levied at the end of the campaign.[46] Livy also explicitly records the *tributum* being levied after the return of the army in 293.[47] Administratively, this also reduced the burden for the state. All that was required of centralised bureaucracy was the total expenditure for the year; no estimates were needed. Thus this element of the *tributum* reconstruction can be accepted.

How, then, were the *tribuni aerarii* to know how much to collect if the amount was not agreed until the end of the year? Past experience would doubtless be a guide, but the issue of accuracy becomes pressing. Collect too much and negative pressure would be put on the patronage networks; too little and the *tribuni* might have to make up the shortfall themselves. Rosenstein suggests that the effort-to-reward ratio was such that *tribuni aerarii* may simply not have bothered to collect *tributum* from the large number of *assidui* who owed very little.[48] On this hypothesis, the importance of accuracy is diminished. On the other hand, Nicolet uses Polybius' disapproval of Roman tightfistedness to suggest that the *tribuni* would have endeavoured to recoup any costs incurred.[49] The reality is probably somewhere between the two. Embedding *tributum* into existing networks of exchange means that the lowest taxpayers would necessarily still be contributing, but it is also unlikely that such rich men as the *tribuni* would regularly expend the effort required to regain (to them) minuscule amounts from a few specific households.[50]

That the right of *pignoriscapio* could apply to a soldier attempting to collect his pay may provide an indication of the efficacy of the *tribuni aerarii* in collecting *tributum*.[51] The requirement for the legislation indicates that the system was imperfect. It is only possible to speculate on the reasons for this: abuse by *tribuni* who deliberately withheld funds or

---

46 Livy 2.54.1, 8.2.4, 9.43.6, 9.43.21, 10.5.12, 10.37.5, 23.43.6, 29.3.5, 30.16.12.
47 Livy 10.46.6.
48 Rosenstein (2016b) 91.
49 Nicolet (1980) 164 with Polybius 31.26–7 – 'no one ever readily and of his own freewill gives away anything of his own to anybody'. See Eckstein (1995) 79–81 on Polybius' disapproval.
50 In such a case, households unable to take part (financially at least) in these networks of exchange were close to the edge. Deliberately pushing them over would have had little short-term financial or social benefit for the *tribuni aerarii*, and possibly longer-term detriment.
51 Gaius, *Inst.* 4.26; Aul. Gell. *NA* 6.10.2–3.

liquidity problems due a lack of coinage, especially in the fourth century, could both be the cause. However, it is worth considering that inefficiency in collection may also have played a part. Relying on integrated networks to collect *tributum* could not have been as accurate as demanding a single set sum from each household. It is plausible that in some years some *tribuni aerarii* did not collect enough to cover the *stipendium*, especially if campaigns were longer or more expensive than anticipated or if farming yields were low. *Pignoriscapio* ensured that the soldiers were still paid, with the *tribuni aerarii* forced to make up any shortfall themselves.

An alternative method of examining the collection of *tributum* is to investigate the occasions on which the system worked in reverse: tax refunds. Examples of this in practice are few. C. Fabricius Luscinus boasted of having refunded *tributum* in 282, and proconsul Cn. Manlius Vulso did likewise in 187.[52] The payment to the people by C. Duilius in 260 may also be an example of a *tributum* refund.[53] Nevertheless, these examples have led many scholars to conclude that *tributum* was, at least in theory, a compulsory loan rather than a tax.[54] This conclusion is based largely on the form of words used to describe Vulso's reimbursement. Livy states that the *tributum* 'which had not been repaid before was repaid'.[55] It has been inferred from Livy's wording that this sort of repayment was regularly expected, although rarely if ever actually received. However, the evidence is not strong enough to support this conclusion. As Nicolet himself notes, the lack of examples is problematic;[56] the refunds were noteworthy to ancient writers precisely because they were unusual. It cannot be assumed that other notices have not been lost, particularly from the lacunae in Livy, but this alone is not enough to prove *tributum* was a loan. Further, while the refund appears to have been in full in 282, the evidence is not sufficient to judge for 260, and the refund was only 2.55% in 187. This is hardly indicative of a system of regular refunds.

What, then, was partially refunded in 187? Livy's use of *stipendium* means that it must refer to *tributum*. Nicolet demonstrates that it is unrelated to the extraordinary loans made in 210 because they had already been repaid, in part with the aforementioned arrears from the priests and augurs.[57] Marchetti suggests that it was an attempt to repay some of the

---

52 Dion. Hal. *Ant. Rom.* 19.16.3; Livy 39.7.5.

53 *CIL* 6.31591 with Rosenstein (2016b) 82. On the inscription see: Östenberg (2009) 58–9; Rowan (2013) 372.

54 Nicolet (1976) 19–26; Nicolet (1980) 156; Ñaco del Hoyo (2011) 382; Rosenstein (2011) 137; Rowan (2013) 374 (tentatively); Tan (2015) 211; Rosenstein (2016b) 82; Tan (2017) xix; Taylor (2017) 159n.2.

55 Livy 39.7.5 – *stipendium collatum a populo in publicam quod eius solutum antea non esset solueretur.*

56 Nicolet (1976) 25.

57 Livy 26.35–6; 29.16.3; 31.13.3–9; 33.42.4; Nicolet (1976) 24; Briscoe (2008) 228.

double *tributum* which had been levied repeatedly during the Hannibalic War.[58] This is plausible, but on the other hand there is nothing in the sources to suggest that this was considered any different to regular *tributum* in terms of state obligation.

It is perhaps the context of this refund which sheds the most light. In 187, Cn. Manlius Vulso was hoping for a triumph. He deposited his spoils in the treasury, as was usual practice, and was only later persuaded by his friends to use some of this cash for public disbursement in order to boost his popularity.[59] Moreover, the senate was not convinced about his right to a triumph; he was accused of having acted as a law unto himself in Galatia with no regard for the Roman state.[60] Pelikan Pittenger highlights that part of Vulso's defence was to suggest that he succeeded because the gods willed it, thus making his victory Rome's. He could only take credit for the victory by sharing it.[61] Thus, the *tributum* refund sits within an attempt to curry favour with both the senate and the people. Refunding *tributum* was a legitimate way to provide a handout; Fabricius and Duilius were precedents. The very vagueness of Livy's reference to unpaid sums may be because it refers not to any one thing in particular but rather more generally to the ideal of war being a benefit rather than a cost. This suggests that all the examples of *tributum* reimbursement are better understood not as loan repayments but as tax rebates.

As modern scholars have noted, there seems to have been an expectation that campaigns should finance themselves.[62] (Whether in fact they were self-sustaining is open to question, but beyond the scope here.[63]) This expectation is reflected in the collection patterns of *tributum*. Examples of campaign years in which *tributum* was not levied are far more frequent than examples of refunds or repayments. In many of these years, the fruits of war, usually in the form of the indemnities, covered the *stipendia* costs. Crucially, Livy frequently records that indemnities were provided specifically to match these costs, sometimes including food and clothing as well as coin for pay. The correspondence with military pay and its deductions cannot be ignored. For example, in 341 the Samnites provided pay for a year and three months' rations; in 325–4 the rations were replaced with a garment for each soldier.[64] In 293 there was uproar in Rome due to the

---

58 Marchetti (1977) 130–1; Marchetti (1978) 263.

59 Livy 39.7.4–5.

60 Livy 38.45.4–7.

61 Pelikan Pittenger (2008) 219.

62 Boren (1983) 430; Serrati (2007) 488.

63 Frank (1933) 95, 145, Rosenstein (2016a) 123, Rosenstein (2016b) 81 and Taylor (2017) 170 have all demonstrated that the massive indemnity payments from Carthage, Philip V, Nabis of Sparta, Antiochus and the Aetolians are all that balanced the books for Mid-Republican military finance.

64 Livy 8.2.4, 8.36.11.

different approaches of the consuls. Both had successful campaigns, but while Sp. Carvilius Maximus forced the enemy to cover the *stipendium* costs, L. Papirius Cursor did not. To make matters worse, Papirius' spoils could have covered the *stipendium*, but he chose to use them differently and so a *tributum* was levied.[65] This demonstrates that there were expectations from the Roman people both that *tributum* was only to be collected when it was financially necessary and that generals should make an effort to limit the burden on the people. Again, this best suits an understanding of the 'refunds' as rebates.

This approach to military spoils, *stipendium* and *tributum* continued into the second century. The trial of the Scipiones was based in part around L. Cornelius Scipio's decision in 189 to award his men rations and double pay with the indemnity from Antiochus.[66] In addition to refunding part of the *tributum*, Vulso also awarded his men double pay. Despite the changing nature of the indemnities, which reached colossal sizes, the link with *stipendium* and *tributum* was not lost. Indeed, if anything, it remained as strong as ever: the collection of *tributum* itself was halted for over a century following the incredible amounts won by Aemelius Paulus in Macedon in 168.[67] As Rowan notes, *tributum* ended only when war booty was large enough to sustain Rome's efforts.[68] *Tributum* began as, and remained until its cessation, an extraordinary tax for a well-defined purpose.

However, the discretion of commanders demonstrated by Papirius indicates that there were other considerations when it came to collecting *tributum*. This is illustrated by the decisions of the Scipiones in Spain. Only twice during the Hannibalic War was money for pay exacted from the province. In the first instance, in 215, the Scipio brothers wrote to the senate asking for grain, clothes and money, but noted that they could raise pay from the province if the senate was able to provide the other necessaries.[69] This raises two points. Firstly, the Scipiones were, unsurprisingly, aware of the financial difficulties facing Rome in 215. Secondly, they knew they could raise sufficient funds in Spain. In turn, this raises the question of why, given Rome's ongoing cash-flow difficulties, this resource had not been tapped previously. Eckstein suggests that the answer is political. The situation in Spain was fluid, and it was unwise to place undue pressure on potentially fragile alliances by asking too much.[70] Necessity forced their hand in 215. This appears to be confirmed by the only other exaction from the war. In 206, Scipio (later Africanus) faced a mutiny. With his troops

---

65 Livy 10.46.1–6, 12–14. For a general's use of war booty, see Shatzman (1972); Rosenstein (2011); Rowan (2013); Tan (2013); Kay (2014) 31–41.
66 Livy 37.59.6. On the trial of the Scipiones see Polybius 23.14; Livy 38.55.5–7.
67 Plutarch, *Aem. Paul.* 38.
68 Rowan (2013) 375.
69 Livy 23.48.4–5.
70 Eckstein (1987) 201–2.

demanding back pay, Scipio had little choice but to source funds on the spot as quickly as possible.[71] Thus it was only in extremis that the Spanish legions relied on the province to provide their pay; local political concerns overrode any desire to reduce the *tributum* burden on Roman citizens.

The discussion can now return to the more practical collection mechanisms. As mentioned, the reimbursement of *tributum* indicates that the system could operate in reverse. Indeed, that the 167 rebate required 2.55% to be returned suggests that this reimbursement could be tightly calculated and controlled. How, though, did this work in practice? At the highest level, it would be a simple matter to return the correct sums to the *tribuni aerarii*. This required only the existing record of what each tribe was required to collect and the tribes' own knowledge of how much each *tribunus* had contributed. The rebate amount could then be calculated (if necessary) and the funds handed over. Taylor suggests that these reimbursements were in fact a credit note, effectively remitting a future *tributum* payment rather than actually refunding that already paid.[72] While the idea of such a note on the census of list of taxpayers is not implausible, given the reconstruction of administration proposed here (see II), the evidence does not support such a conclusion in this case. As Nicolet highlights, the language used by Livy for the Vulso example repayment is exact; if he had meant remittance, *remittere*, not *solvere*, would have been used.[73] Another explanation is required.

It is at this point that the matter becomes murkier. The embedded social and financial networks could work effectively to gather *tributum* across a year, but it is less clear how these systems could do the opposite, especially to the accuracy suggested by 2.55%. Unfortunately, the complete silence of the sources on this point means that nothing more than conjecture is available. Theoretically, if *tribuni aerarii* were able to calculate who owed what, the reverse should also be true. Perhaps the same networks were used to reimburse individual households. Alternatively, especially in the lower census classes, perhaps the rebates were felt more as remittances in the following years; it may have been simpler than ensuring that the correct cash filtered all the way back through a system which operated in kind. Whilst these were refunds from the state's perspective, it does not mean that they remained so further down the social scale. Nonetheless, the political cachet associated with a *tributum* rebate indicates that they were believed by both generals and the people (or at least those politically active people in Rome) to have an influence on ordinary citizens, even if in reality some may have barely felt it.

In summary then, the mechanisms to collect *tributum* are almost invisible. *Tribuni aerarii* were the linchpins in large social and financial

---

71 Polybius 11.25–30; Livy 28.24.1–29.12; Appian, *Hisp.* 34.137–36.146.
72 Taylor (2017) 160.
73 Livy 39.7.5; Nicolet (1976) 25n.45.

networks, through which they were ideally placed to collect *tributum* contributions on behalf of the state, as reflected by their continuing status after *tributum* was halted. This ability hints at the hidden importance of the tribe within Roman society, and provides a brief snapshot of a non-magisterial elite. Tribal mechanisms, whether in some centralised fashion or devolved again to the *tribuni aerarii*, were able to provide the more complex, on-the-ground organisation that Rome's small central state was simply not equipped to perform. As in other areas of Mid-Republican military administration, complexity can be glimpsed, but unlike elsewhere there is simply not enough evidence to say much more.

iv Paying *stipendium*

The final element of the *tributum* system to be examined is how it was paid to the soldiers as *stipendium*. This section discusses the process in order to demonstrate that this was one area in which the *tributum* system was altered to suit the changing circumstances of Rome's longer, more distant wars.

Varro provides evidence for how *stipendium* was originally paid: the *tribuni aerarii* who collected it also paid each soldier individually.[74] The *tribuni aerarii* who appear in Plautus' plays demonstrate that soldiers were required to seek out the *tribuni* for themselves, as does the legal redress of *pignoriscapio*.[75] Just as with the collection of *tributum*, there was no centralised mechanism for the distribution of pay. Instead, Rome remained reliant on local networks and local actors to meet *stipendium* requirements. This raises an interesting question: how did the *tribuni aerarii* know what was owed to each individual soldier? The previous chapter demonstrated that detailed records of individual pay deductions were kept by the quaestor (3:iii), but how did this knowledge reach the *tribuni*? It is implausible that a system which functioned with so little central oversight suddenly required the large-scale transfer of military records at this point. Perhaps the best evidence is from Plautus. A veteran is able to extort money on the pretence that the victim is a *tribunus*.[76] This suggests that it was the soldiers who requested the amount, perhaps presenting their own pay records. Such a system appears open to abuse, but the same networks which collected *tributum* may have limited this. The soldiers were part of these networks, and on discharge were about to return to them. That most citizens served meant that 'average' pay levels would have been common knowledge. Social pressures may well have limited an individual's attempt to claim more.

74 Varro, *Ling. Lat.* 5.181.
75 Plautus, *Poenulus* 1285–6; *Aurularia* 525ff; Gaius, *Inst.* 4.26.
76 Plautus, *Poenulus* 1285–6 – *aere militari tetigero lenunculum*.

Alternatively, the provision of *stipendium* by *tribuni aerarii* may provide evidence for the inception of detailed pay records kept on campaign. If, as suggested, widespread abuse was prevented by the social fabric in which *tributum* operated, detailed individual pay records may not have been required. Local, smaller campaigns would have costs that could be easily estimated. Further, there is no place in this distribution system for the deductions described by Polybius.[77] The central state had no involvement. Instead, this system reflects much more the original intention of *stipendium* as described by Livy: the individual's war costs covered by the community.[78] Prior to Veii, men had paid for their own supplies. There is no evidence that the provisions for these short, local campaigns were organised centrally by the general or his quaestor. Thus, the only change brought about by the introduction of *stipendium* was that men were reimbursed for their costs at the end of the campaign. It is probable, therefore, that the development of the more complex pay records found in Polybius ran in parallel with the centralisation of *stipendium* (see later). Exactly when this occurred is uncertain, but sometime by the mid-third century would match the geographical development of Rome's warfare.

That a major change in the method of *stipendium* distribution occurred during this period is demonstrated by Polybius. He states that by the time of the Hannibalic War, it was the quaestor on campaign who was responsible for providing pay.[79] Polybius is explicit, and modern scholars have found little reason to doubt him. As discussed previously (I:ii), Polybius is an authority, and his military digression can be accepted as accurate. Further, almost all other references to pay distribution in the Middle Republic place the quaestor in this role.[80] The need for this development, from pay in Rome from the *tribuni aerarii* to pay on campaign from the quaestor, is evident from the sources. The *tributum* distribution system suggested by Varro and Plautus was no longer fit for purpose once longer, farther-flung campaigns became a regular feature of Roman warfare. As *tributum* levels were set for a given year, it follows that the proceeds were paid out likewise. There is no evidence that serving men were given yearly furlough to return home to collect pay, and it is implausible that the *tribuni aerarii* (the number of which Tan estimates at around 5000 in 225)[81] went to the legions. Legions were enrolled by tribe in a manner which split members across several legions. The men that a single *tribunus* might have to pay could be spread across the Mediterranean. Nor is this just a Mid-Republican problem; garrisons were

---

77 Polybius 6.39.15; see III:iii.
78 Livy 4.59.13.
79 Polybius 6.39.15.
80 See following discussion for references.
81 Tan (forthcoming).

overwintering as early as 342,[82] meaning that the limitations of Varro's distribution system must soon have been felt.

Nonetheless, before continuing there is an objection which must be overcome. If the *tribuni aerarii* had lost their payment role by the late third century, why does second-century Plautus still use this image? Indeed, Varro quotes Plautus as evidence.[83] Hill suggests that Plautus is referring to an institution threatened with extinction, but according to Polybius this role had already been lost.[84] Plautus' *tribuni aerarii* are perhaps best understood as cultural ghosts, much as the traditional image of the bank manager is today. The role and title still existed (there is no evidence that their role in tax collection disappeared until the tax did), but the *tribuni* themselves were no longer involved in direct contact with serving soldiers.

Before he could pay the men, however, the quaestor first had to receive the *stipendium* funds. The clearest example of this is provided by Sallust. During the Jugurthine War, Sallust describes the urban quaestor C. Octavius Ruso transporting the *stipendium* to Marius in Africa.[85] Likewise, Cicero recounts Verres as quaestor transporting money to his consul in 84.[86] These examples date from after the period under discussion, but it is difficult to see this as anything other than the usual process.[87] In 180, Fulvius Flaccus boasted of having funded the *stipendium* from his province;[88] likewise, in 215 the Scipio brothers were also able to fund the *stipendium* locally, albeit under very different circumstances.[89] The implication is that it was normally sent from Rome, and the context in 215 demonstrates that this money came from the treasury in particular. The unusual case of Ligurian legions, also in 180, provides more direct evidence. When these men were dismissed early by the military tribune M. Fulvius Nobilior, the centurions were ordered to hand over to the *aerarium* (treasury) any funds they held.[90] Boren argues that this is evidence of men receiving pay advances which, due to the circumstances, had to be repaid.[91] More importantly here, it demonstrates that the treasury was operating as a central financial repository for all military finance. Not only were spoils deposited there, *stipendium* was too.

82 Livy 7.38.4.
83 Varro, *Ling. Lat.* 5.181.
84 Hill (1946) 61.
85 Sallust, *Iug.* 14.3 – *stipendium in Africam portauerat*; Pina Polo & Díaz Fernández (2019) 97.
86 Cicero, *Verr.* 2.1.14.36.
87 Boren (1983) 434.
88 Livy 40.35.4.
89 Livy 23.48.5.
90 Livy 40.41.8.
91 Boren (1983) 434.

This, then, was the only direct change to the *tributum* system required by the change from short, local campaigns to long, overseas wars: rather than paying the men directly, it appears that *tribuni aerarii* were required to deposit the *tributum* with the treasury. *Tribuni aerarii* never were and never became treasury officials (unlike some *scribae*, for example; VI:ii), and there is no evidence that the *aerarium* took on a role in local tax collection. However, when the logistics of Rome's military endeavours forced a degree of centralisation, it is unsurprising that the *aerarium* became that central location. The treasury was the natural home of money and had existing links to the military and its administration.[92] Moreover, it already had financial officials on the ground with the legions, in the form of the quaestors, whilst the urban quaestors were present in Rome to collect the contributions from the *tribuni aerarii*. Rome made use of existing institutions, limiting change whilst adapting to new demands.

Such a system, where the quaestor collected the *tributum/stipendium* from the treasury in Rome at the end of the campaign year, is entirely in keeping with both the wider military administration proposed in this monograph and the tendency to keep the involvement of central institutions to a minimum. The Scipio brothers demonstrate that the request for pay could be made in writing as well as in person. Previous chapters have shown that letters between the legions and Rome were relatively frequent and provided reasonable detail. If necessary, quaestors could calculate the funds required from their campaign records; the treasury required only the final figure. However, it is unlikely that even this calculation was required. Men had deductions made for their expenses, but the legion still had to meet these costs. Commanders probably requested the full *stipendium* amount. The minimal pay grades of the army made calculation a simple task of addition and multiplication.

v Quaestorial accounts

A discussion of *tributum* and *stipendium* in the context of military administration would not be complete without discussing the accounts which recorded these funds. The minutiae of individual pay records have been discussed already (III:iii), but macroeconomic records were also required in order to calculate the *stipendium* and note its use. One lone set of quaestorial accounts survives from the Republic: those of Verres, who served as quaestor to Cn. Papirius Carbo in 84–2.[93] These accounts only survive at all due to a set of extremely unusual circumstances: they were included by Cicero, apparently in full, in his prosecution of Verres. Whilst the first century is technically outside the scope of this Mid-Republican study, it would be negligent to ignore this evidence entirely. An examination of both Verres and his accounts yields some useful insights into quaestorial accounts and the wider role of the quaestor.

---

92 For a detailed discussion of the *aerarium Saturni*, see V:iii.
93 Cicero, *Verr.* 2.1.14.36.

The office of quaestor had been, since its inception, an administrative position linked to the use of state finances.[94] Within the legion, the quaestor had a significant role in both administrative and financial matters. Roth argues that the space Polybius allotted to the quaestor in the camp reflects the number of men and wagons beyond the legion proper whom the quaestor commanded in order to carry out his duties,[95] suggesting that the quaestor had control of the baggage train which accompanied Mid-Republican armies. However, Polybius does not mention the quaestor in his account of legion recruitment, indicating that he and his staff lay slightly outside the hierarchy despite their integral role. This may in part be due to the fact that the quaestor was, by the third century, elected, and his *provincia* then allotted as with the consuls.[96] The quaestor functioned as a representative of the senate and the treasury. It was to the quaestor that the funds for a campaign were released, and by the quaestor that records of spending were kept.[97] The quaestor as well as the commander kept financial records, indicating that part of the quaestor's role was to be a check against corruption.[98] The scrutiny under which these records could come only emphasises both the importance of accounting and the potential for tension between commander and quaestor.[99]

The quaestor's role as a senatorial representative to some degree independent of the general could cause friction. The attitude of Verres' quaestors in the first century is instructive here. Cicero was given charge of the prosecution of Verres over Verres' own quaestor. The quaestor wished to act as prosecutor due to the personal injury he had received.[100] The exact harm is not specified, but it can be inferred that the injury was due to a manipulation of accounts in the face of the quaestor. Interestingly, Cicero later goes on to complain that the quaestors stationed in Sicily hindered his investigation due to loyalty to Verres.[101] For Verres' corruption and embezzlement to have been as successful as Cicero argued, the cooperation of the quaestors who managed the parallel accounts was required.

This points to the method by which these potential tensions could be relieved: a 'special relationship' between general and quaestor.[102] Once again, Verres provides a good example. Cicero accused Verres of

94 Cf. Tacitus, *Ann.* 11.22.4–6 (*ut rem militarem comitarentur*); Mommsen (1894) 223–9; Latte (1936) 24–33; Harris (1976) 92–106. For a full discussion of the origins of the quaestorship, see Pina Polo & Díaz Fernández (2019) 5–24.

95 Roth (1999) 258; Polybius 6.32.8.

96 Tacitus, *Ann.* 11.22.4–6.

97 Cicero, *Verr.* 2.1.14.36.

98 Daly (2002) 122; Pina Polo & Díaz Fernández (2019) 168.

99 Rosenstein (2004) 253 n.9; Plutarch, *Ti. Gracch* 7.1–2; Livy 30.38.1; Thompson (1962) 353; Cicero, *Fam.* 5.20.

100 Cicero, *Verr.* 2.1.6.15 – *ut ei qui istius quaestor fuisset, et ab isto laesus inimicitias iustas persequeretur.*

101 Cicero, *Verr.* 2.2.4.11.

102 Mommsen (1894) 266.

embezzling from army funds whilst serving as quaestor to Carbo. Cicero's attack focuses as much on the 'violation of the personal tie imposed and sanctified by lot' as the theft itself.[103] This indicates that the appointment as Carbo's quaestor created not only a state obligation but a personal relationship to be valued by both parties. It suggests that, as Cicero states, abandoning the personal tie was tantamount to deserting the army as a whole.[104] Maintaining the relationship overrode and smoothed any potential tension from the quaestor's more ambiguous position in the legion. The defence of Verres' quaestors for his abuses in Sicily is a demonstration of this relationship in action, although with the opposite outcome to that desired by the state.

The personal relationship between general and quaestor can also be seen in the Middle Republic. When Cato the Elder served as Scipio's quaestor in 204, the two clashed over the use of funds.[105] However, this conflict should not be overstated. Cato is recorded as complaining about Scipio, but there is no evidence that he more actively attempted to prevent Scipio's spending. This seems to be the 'special relationship' in action; however, although the quaestor was a senatorial representative, he was nonetheless an inferior magistrate under his commander's *imperium*. Indeed, Polybius states that quaestors faithfully obeyed their consuls.[106] There is no need to retroject later antagonism between Cato and Scipio to 204 in order to explain this relatively minor incident, although it may well be why ancient historians considered it noteworthy. Rather, it may be best to see Cato's complaint as a demonstration of the quaestor's role and the limits of his power over his commander.

As Roth highlights, the quaestor's role was primarily an administrative one, but he could take over were the commander of the legion killed.[107] Again, this should not be surprising. In order to become eligible to stand for the magistracy, a quaestor had (in theory) to serve ten years/campaigns in the army.[108] With the possible exception of the senior military tribunes (who had served for ten years and were occasionally consulars[109]) and any lower-ranking career soldiers, the quaestor was the most experienced officer

---

103 Cicero, *Verr.* 1.4.11 – *sortis necessitudinem religionemque violatam*. Verres was apparently able to get away with this theft despite two sets of books because he was also keeping the consul's on his behalf (2.1.14.37).

104 Cicero, *Verr.* 1.4.11.

105 Plutarch, *Cato Mai.* 3.5–6.

106 Polybius 6.12.15.

107 Roth (1999) 258; Appian, *Hisp.* 11.63.

108 Polybius 6.19.4.

109 Polybius 6.19.1. In 216, Livy (22.49.16–17) reports more than 80 senators dying at Cannae, including consulars. It must be assumed that these men were holding positions such as military tribune. While Cannae must always be considered an abnormal example, it may nonetheless indicate that at least in the third century it was not unusual for holders of high magistracies to serve again in less senior positions.

on the ground, and the only one holding a magistracy. Even if his exact position in the hierarchy ran parallel to the legion proper, the quaestor would be the obvious candidate to assume control.[110]

It is within this context of consul–quaestor tension that the single surviving set of quaestorial accounts appears. Cicero reads them, apparently in full, in his prosecution of Verres for his actions in Sicily as consul. To support his case, Cicero also accuses Verres of embezzling previously whilst he was Carbo's quaestor. Cicero produces the campaign's financial records to support the allegation:

> '*Accepi*' inquit '*viciens ducenta triginta quinque milia quadringentos X. et VII. nummos. Dedi stipendio, frumento, legatis, pro quaestore, cohorti praetoriae HS mille sescenta triginta quinque milia quadrigentos XVII. nummos. Reliqui Armini HS sescenta milia*'.
>
> (Cicero, In Verrem 2.1.14.36)

> It said 'I received 2,235,417 HS. I gave for pay, grain, legates, for the quaestor and for the praetorian cohort 1,635,417 HS. I left at Arminium 600,000 HS'.

The passage reveals several useful facets of the quaestor's role and his accounts. Cicero chose to quote this account due to his incredulity at its form.[111] The implication is that Verres should have included much more detail. In particular, Cicero compared Verres' accounts with those of P. Servilius Vatia Isauricus, who provided detail on the size and nature of the booty acquired.[112] Scipio Africanus' quaestor C. Flaminius can also be found creating an inventory of captured material in New Carthage in 209.[113] Pina Polo and Díaz Fernández suggest that Livy's triumphal booty records may originate from inventories created by quaestors in the field and later stored in the treasury.[114] This indicates the type of information to be included in quaestorial accounts and its future uses. However, unfortunately, Cicero provides no further detail about what else might be expected. Was prose the usual form, for example, or would something akin to single-entry book-keeping be more the norm? Cicero either expects his audience to know how proper accounts should appear or does not consider it relevant to his case. Nevertheless, it can be concluded that the quaestor's accounts should have been a great deal more detailed.

---

110 Mommsen (1894) 269 argues that the quaestor was the most senior officer after the general, based on the number of guards placed by his tent (Polybius 6.35.4).
111 Cicero, *Verr.* 2.1.14.36.
112 Cicero, *Verr.* 2.1.57.
113 Livy 26.47.8 with Pina Polo & Díaz Fernández (2019) 177.
114 Pina Polo & Díaz Fernández (2019) 171.

Despite this lack of specificity regarding proper form, the passage reveals and confirms other details. Firstly, the list of items to be paid for by the quaestor agrees with that given by Polybius in his discussion of military matters.[115] This suggests that, while conclusions drawn from first-century evidence do not always apply equally to the third and second, the quaestor's role was similar, and it is not unreasonable to test such conclusions against the Middle Republic. Secondly, despite the account's brevity it does contain specific figures. The remaining figure of 600,000 HS appears suspiciously round. Scheidel has argued that figures of this type can simply mean 'a lot' rather than reflecting a real number.[116] Scheidel has also pointed out that Cicero changes his mind about exactly how much he is accusing Verres of embezzling in total.[117] However, as Cicero claims to be quoting the accounts here, the figure of 600,000 HS can be broadly accepted, even if its exactness remains in doubt. More importantly, the initial amount quoted by Cicero seems to be exactly what the treasury provided Verres. This has two implications: that this figure could be checked against treasury records to demonstrate its accuracy; and that the quaestor was expected to deal in specific values, not estimates or round figures. As the Mid-Republican quaestor performed the same role, it is not unreasonable to suggest that the same, or at least similar, standards applied. This is reflected in the attention to detail found in the individual pay records discussed previously (III:iii). The *stipendium* collected from the treasury by Verres was not derived from citizen *tributum* (as its collection had ceased in 167), but there is no reason to believe that accounting practices, and the systems behind them, were not otherwise the same.

vi *Tributum, stipendium* and military administration

As with all military administration discussed in this monograph, Rome made the minimum necessary changes to the *tributum/stipendium* system to meet the changing demands of its growing ambitions. The practical change of having *tribuni aerarii* pay the *tributum* to the treasury rather than directly to the soldiers was small, entirely practical and indeed unavoidable if Rome was to continue to pay its men. Conceptually, however, it was huge.

In the fourth century, once censors had assigned census ratings to households, the central institutions only had to provide every tribe each year with a figure for its tax liability. Beyond that, the collection of *tributum* and distribution of *stipendium* were left to the *tribuni aerarii* and completely outside central state control. It cannot be overstated how

---

115 Polybius 6.39.15.
116 Schiedel (1996) 222–38.
117 Scheidel (1996) 228 with Cicero, *Verr.* 1.56, 2.1.27, 2.2.26 (40,000,000 HS) and Cicero, *Q. Caec.* 19 (100,000,00 HS).

remarkable it is that Rome was able to successfully fund its expanding military with almost no central oversight. This all changed once longer, overseas campaigns became the norm. At some point in the first half of the third century, the central state's role went from almost entirely hands-off to key coordinator. *Stipendium* distribution now required central organisation. Moreover, this centralisation had ramifications on broader financial operations too. Payment through the treasury suddenly gave Rome the resources to centralise other military matters, such as procurement. As wars lasted longer and were further from home, centralising procurement helped facilitate their successful operation, as reflected by the use of *publicani* and the collection of tithes in kind. The result of this was that the state too now needed to be reimbursed. In combination, these factors changed military pay from the strict reimbursement of outlay described by Livy to the more complicated system of deductions familiar to Polybius, prompting the development of the administrative systems seen here and in previous chapters.

However, there were limits to the changes in *tributum* brought about by Rome's growth. The purpose of the tax – to fund *stipendium* – remained fixed from its inception in the early fourth century to its cessation in 167. While the scale of Rome's operations turned the irregular tax into a yearly one functionally, conceptually it had not changed. Even under times of extreme duress, there was no attempt to broaden its purpose or alter the census ratings upon which it was based in order to better access the growing private wealth which accompanied Rome's expansion. That *tributum* ceased to be levied after Aemilius Paulus' victories in 167 meant Rome sidestepped the inherent tension in the system between the quantity of manpower and the number of taxpayers to fund them. Given the recruitment and agrarian problems of the later second century, it is questionable how much longer the system could have survived unchanged. This tension was only ever patched over, but for the Middle Republic patches were more or less sufficient. Nor were *tributum* collection mechanisms altered, leaving the legacy of the status of the *tribuni aerarii* long after their practical role had ceased. As in other areas of military administration, the changes made were only those forced by absolute necessity.

In establishing the nature and complexity of the paperwork connected to the Mid-Republican legions, individuals responsible for creating and keeping these documents have been frequently alluded to or their existence assumed. Having established the kind of records which allowed Rome's armies to operate – *tabulae iuniorum* and legion lists with details of individual service – it must be questioned who these bureaucrats were, if indeed such a term can be used. To complete the picture of Mid-Republican military administration, these individuals on campaign and in Rome must be identified, as well as how the documents were stored and accessed.

5   Documents and archives

The preceding chapters have established that the Roman Republican army was administratively able; it operated with a considerable degree of bureaucracy both in the field and in Rome. This chapter examines the practical issues of record keeping: the physical materials, their size and where they were stored in Rome. A lack of direct evidence means that the solutions to these problems are largely hypothetical. The conclusions drawn focus on the balance of probability and plausibility in order to gain a sense of the physical reality of record keeping.

i Physical form of records

This section attempts to identify the materials used for Roman army records. Pliny the Elder provides a passage on the use of writing materials in his *Natural History*:

> *Prius tamen quam degrediamur ab Aegypto et papyri natura dicetur, cum chartae usu maxime humanitas vitae constet, certe memoria. et hanc Alexandri Magni victoria repertam auctor est M. Varro, condita in Aegypto Alexandria; antea non fuisse chartarum usum. in palmarum foliis primo scriptitatum, dein quarundam arborum libris, postea publica monumenta plumbeis voluminibus, mox et privata linteis confici coepta aut ceris: pugillarium enim usum fuisse etiam ante Troiana tempora invenimus apud Homerum.*
>
> (Pliny, HN 13.68–9)

However, before we move away from Egypt, the nature of papyri shall be discussed as well, since the civilisation of life, certainly memory, depends greatly on the use of paper. And Marcus Varro reports that this was made known by the victory of Alexander the Great and the founding of Alexandria in Egypt; before which there was no use of paper. First writing was on the leaves of palms, then on strips of certain trees, afterwards for public records on lead sheets, and soon they began to use linen or wax for private documents; for we find in Homer that little wooden books were used even before the Trojan era.

Bronze

Pliny's discussion, however, appears incomplete. It does not mention the most securely attested medium for ancient records: bronze.[1] Bronze records were used throughout the Roman period. When fire damaged the Capitol in AD 69, Suetonius reports that Vespasian endeavoured to recreate more than 3000 melted inscriptions.[2] Vespasian (or Suetonius) believed that these inscriptions dated as far back as the foundation of Rome. Polybius describes a series of treaties recorded on bronze between Carthage and Rome dating from the beginning of the Republic to the Second Punic War. He also believed in their authenticity, mentioning universal difficulty in reading their archaic Latin. The tentative dating of *CIL* 1^2 2833 to the regnal period suggests that bronze inscriptions could have been produced very early in Rome's history (although the inscription itself is not from Rome). None of this is indisputable evidence, but it does demonstrate a belief that Rome had maintained an epigraphic habit throughout its existence in order to record important decisions.[3]

Pliny's omission of bronze, however, is directly related to its use. Vespasian was searching for copies of *senatus consulta*, laws, alliances, treaties and special grants of privilege to individuals.[4] The three treaties described by Polybius were likewise significant public events.[5] Polybius included these treaties in his work explicitly because others were unaware of them, perhaps suggesting they were in the temple rather on display. However, this need not suggest archival storage. They may well have been taken down following the Second Punic War, due to the new state of Romano-Punic relations. The evidence for bronze inscriptions is thus monumental, not archival.[6] This monumental nature may be why Pliny excludes bronze from his description of writing materials.

Papyrus

Pliny indicates that papyrus was the primary writing material for storing public documents in his time. The term *charta* appears to be

---

1 For example, the previously discussed *tabula Heracleensis* and the *Lex Ursonensis* are both bronze inscriptions.

2 Suetonius, *Vesp.* 8.5.

3 Cf. Cornell (1991); Langslow (2013) 176–8.

4 It is highly unlikely that *senatus consulta* were regularly recorded on bronze. Many were in response to immediate, time-limited problems like yearly legion recruitment and deployment; such *consulta* would be effectively out-of-date before they could be displayed in bronze.

5 Polybius 3.21–6.

6 Frederiksen (1965) 186; Edmondson (1993) 162–3; contra Bucher (1987) 6. Williamson (1987) 162–3 discusses the practical difficulties of consulting bronze tablets, including legibility and style.

interchangeable with *papyrus*.[7] Papyrus was for durable, long-lasting records, as *certe memoria* confirms. That Cicero kept master copies of his speeches on papyrus rolls (rather than in another form), apparently for longevity, supports this interpretation.[8] However, while papyrus may have been the main material for Imperial document storage, Pliny is clear that it was not the only medium.

When did papyrus become the main administrative material in Rome? Pliny was writing in the late first century AD; it cannot be assumed that his description applies equally to the Middle Republic. Pliny dates papyrus use in the wider Mediterranean to 332; if trustworthy, this provides a possible *terminus post quem*, but it does not indicate whether Rome took up papyrus at the same time. Pliny (and Varro) is wrong that papyrus was unknown outside Egypt before Alexander's conquest: both Herodotus and Demosthenes, for example, mention it.[9] In Rome, Mid-Republican Ennius mentions papyrus in his *Annales*, but he is concerned with the production of literary works, not administrative documents.[10] That third-century Roman poets and historians may have followed the Greek model of producing works on papyrus does not mean that it was also used for everyday administration.

Bucher suggests that widespread Roman use of papyrus for administrative functions should be dated to the early first century. He argues that a change in epigraphic forms from massed text to columns indicates that engravers began copying text provided in columns, as found on papyrus. The lower character density of this column form suggests that the change was not economically driven, and is thus best explained by a change in writing material to papyrus.[11] Bucher's theory is attractive, as it suggests increased adoption of papyrus in everyday administration. This does not rule out the use of papyrus before this point, but it suggests that it was not the Middle Republic's primary writing material. On balance, the evidence for other materials, coupled with the first-century change in epigraphic forms, suggests that papyrus was not widely used in the Middle Republic.[12]

---

7 Cf. *OLD* s.v. *charta*.

8 Cicero, *Q. Fr.* 2.11.4; Nepos, *Atticus* 16.

9 Hdt. 5.58.3; Dem. 56.1.

10 Ennius, *Ann.* 458 Sk. This use further undermines Pliny's strict division between public and private document materials.

11 Bucher (1987) 15–6. Frederiksen (1965) 188 argued this previously in a more abbreviated form.

12 The overall lack of pre-first-century references to papyrus may be due to the period's sources, rather than reflecting a genuine situation, but the weight of other evidence means it is not necessary to rely on an argument from silence.

Leaf-style tablets

Pliny provides several alternatives to papyrus which he believes predate its use. Writing *arborum libris* seems to be a reference to leaf-style wooden tablets, examples of which have been discovered at Vindolanda. These tablets were formed of a thin sheet of wood approximately the size of a postcard. This was scored down the centre to create a fold and written on with ink. Once folded, the document could be sealed with string. Several leaf tablets could be attached to one another for longer documents.[13] These leaf tablets were used for second-century AD military documents, making it tempting to conclude that they had a similar use in the Middle Republic. In concert with Pliny's assertion that wooden tablets predate papyrus, this appears a convincing argument. However, care must be taken. Bucher, following Bowman and Thomas, argues that the leaf tablet was a north-western development which paralleled the use of papyrus in the Mediterranean basin.[14] The leaf tablet provided an easily obtained, cheap, disposable writing surface in an area far from papyrus production.

There are, however, several problems with this argument. Leaf tablets of local wood are attested in the Mediterranean: Herodian describes lime-wood leaf tablets in Rome.[15] If Bowman and Thomas are right to conclude that Herodian added detail to an account originally written by Dio, Herodian's familiarity with the leaf-style tablet indicates that they were not unusual in the southern empire.[16] Likewise, Martial refers directly to this tablet type without reference to the northwestern provinces in particular.[17] In the quoted passage, Pliny attributes leaf tablets to the Trojan period. The *Iliad* is hardly incontrovertible evidence for the 'Trojan period', but it does mention leaf tablets, suggesting that they were familiar to whoever transcribed the poem in c. eighth-century Greece.[18] Further, Bowman and Thomas have convincingly argued that the term *pugillaria* often refers not to wax tablets (as is usually believed) but specifically to the smaller leaf tablets.[19] This does not directly demonstrate Republican use of the leaf tablet, but it does indicate widespread use across space and time. The leaf tablet was not a second-century AD invention.

The long use of leaf-style tablets helps to avoid any problems with Pliny's terminology: *arborum libris* literally translates as 'the bark of trees'. Meyer points out that Nero accepted an allegedly contemporary account of the

---

13 See Bowman & Thomas (1983) 37–9 for full description and diagrams.
14 Bucher (1987) 27; Bowman & Thomas (1983) 44.
15 Herodian 1.17.1.
16 Bowman & Thomas (1983) 41; Dio 67.15.3.
17 Martial 14.3.
18 Homer, *Il.* 6.168.f.
19 Bowman and Thomas (1983) 43. They admit that the evidence is far from conclusive for this but suggest that at the least the term can mean both.

Trojan War due to its material: *tilia* (limewood or bark).[20] Pliny states that the same material was used by army scouts to write reports.[21] As seen, limewood was also used for leaf-style tablets. As well as confirming the long use of this material, the mentions of *tilia* as both limewood specifically and bark generally indicate a blurring of meaning. This allows both *tilia* and *arborum libris* to mean leaf-style tablets as well as bark.

Two other objections to Bucher's conclusions must be dealt with. There is no need to doubt that leaf-style tablets were a relatively cheap material produced locally. This would have made them attractive in most areas of the empire. Transportation from Egypt would have made papyrus that bit more expensive wherever in the empire it was used. There is no evidence that papyrus was considered a particularly cheap or disposable writing material outside Egypt. Therefore, it should be concluded, as Bowman and Thomas lean towards, that leaf-style tablets were a much more common writing material than previously thought. Their fragility means that only specific circumstances allowed them to survive into modern times, and common use meant that ancient writers, even Pliny the Elder, felt no need to explain them.[22] In this vein, it is perhaps best to see not Vindolanda's leaf-style tablets but Egypt's papyrus documents as local peculiarities.

The nature of the military records found at Vindolanda and in Egyptian papyri indicates their intended longevity.[23] Equipment and absentee lists fall into a similar category to the majority of *senatus consulta*: records to be kept for a period, but not for many years.[24] Indeed, these documents only survive into the present day because they were thrown away. This suggests that leaf-style tablets would have formed a good material for the proposed records generated on campaign. Individual service and pay records for an entire legion would be well suited to a small, light, easily produced material. Should additional space be required, another leaf could be easily manufactured and attached. Once crucial elements had been transferred to a military record with the census, their role was completed and they could be disposed of.

The *tesserae* mentioned by Livy and used to pass orders without using trumpets may be examples of Mid-Republican leaf-style tablets.[25] A more specific definition of the term *tessera* is not provided, but the word in this context suggests a small sheet of wood.[26] Had a wax tablet been meant,

20 Meyer (2004) 35; Septimus, *Dictys* Pr.; OLD s.v. *tilia*.
21 Pliny, *HN* 16.35.
22 Bowman & Thomas (1983) 44.
23 See Introduction.
24 *Senatus consulta* were supposed to be kept in the *aerarium*, suggesting that long-term storage was the aim. This need not imply that they were frequently consulted, but it does suggest a greater durability than required by an army on campaign.
25 Livy 7.35.1, 9.32.4, 27.46.1, 28.14.7, 39.30.4.
26 OLD s.v. *tessera*. More broadly the term means a flat piece of material, but Polybius (n. 28) suggests that wood is an appropriate translation here.

tabula would most likely have been used. Harris considers Livy's late-fourth-century examples to be annalistic interpolation.[27] However, there is no reason to follow him other than a general scepticism about accounts of the fourth century, to which the author does not subscribe. Even if Harris is correct, Livy's examples indicate that silent orders were commonly passed on small wooden tablets by the Late Republic. As they were strategically advantageous in the situations Livy narrates, it is plausible that *tesserae* were in use by the late fourth century.

Moreover, Polybius' description of watchkeeping in camp is reliant on wooden writing sheets.[28] On one occasion he refers to this as κάρφος, a term which ordinarily means a small piece of wood or kindling.[29] This indicates that he envisaged the tablets as slivers of wood, not wax tablets.[30] Further, he details actions requiring written instructions which appear to have required more than a few words. Watch inspectors received the inspection order in writing, and the role was reserved for the more literate (see VI:ii). This suggests that the sheets were more than scraps, again suggesting leaf-style tablets. In combination, these passages of Livy and Polybius suggest that leaf-style tablets were commonly used in the field by the late third century at the latest.

Wax tablets

With papyrus little used in the Middle Republic, and leaf-style tablets used in the field but intended for disposal, an alternative is required for document storage in Rome. The other wooden writing material listed by Pliny is the wax tablet. This is the best attested writing material.[31] It was formed of a wooden tablet with a recess. Once the recess was filled with wax, writing was scratched into it with a stylus. These tablets usually came in pairs which could be sealed together with the writing on the inside.

It is assumed from now on that when *tabulae* are mentioned by ancient authors, it is almost certainly a reference to wax tablets. A certain amount of discretion must be exercised when dealing with the term; *tabulae* were not universally wax tablets. For example, the Twelve Tables, *XII Tabulae*, which contained the first written law of Rome were bronze plaques. However, it has already been established that bronze had a monumental use. Pliny avoids the term *tabulae* entirely, instead using the common abbreviation *cerae* for wax tablets. As *pugillaria* appears to be the common

27 Harris (1989) 167 n.92; Livy 7.35.1, 9.32.4.
28 Polybius 6.34.7–12, 6.35.5–36.8.
29 Polybius 6.36.3. LSJ s.v. κάρφος.
30 Polybius (6.34.8) does use the phrase πλατεῖον ἐπιγεγραμμένον, but there is no need to translate this as anything more than 'written-on tablet'. It does not imply the engraving required with wax tablets.
31 See Meyer (2004) 26 for an extensive list of ancient references.

term for leaf tablets, and it will be seen that *libri* is often used for linen, wax tablets are the best candidate for most mentions of *tabulae*. Unless otherwise indicated, *tabulae* will now be interpreted as wax tablets.

As a storage material, wax *tabulae* have several advantages over papyrus and leaf-style tablets. Their sturdiness made them more durable; Bowman and Thomas suggest this is why more wax than leaf-style tablets survive.[32] They were also waterproof, making them ideal for transporting messages. Messages or documents scratched into wax were much more likely to survive for a longer period, and could be trusted to be immune from deliberate or accidental change. By contrast, the ink-written whitewashed boards displayed in the *forum* were notorious for being alterable.[33] Moreover, the most common ink was water soluble.[34] The thickness of wax tablets in comparison to leaf tablets increased their durability in a storeroom's potentially damp confines. Vitruvius instructs that when building a house the *bibliotheca* should be placed facing east to take advantage of the extra light and avoid the dampness which could occur in a westward facing room.[35] As will be seen (V:iii), exactly where military documents were kept is unclear, but the majority of probable locations were enclosed spaces without an eastward outlook. Wax tablets had an advantage in both security and durability over other materials.

Wax tablets do, however, have disadvantages. The bulkiness which renders them more hard-wearing also makes them more difficult to store. Indeed, the main arguments against wax tablets as archival material focus on the difficulties of using wax tablets for long documents and storing the required quantities.[36] This is particularly true of census records. However, as Bucher himself recognises, the late-first-century Ahenobarbus relief is commonly interpreted as depicting the census being recorded on wax tablets.[37] Likewise, the tax records being thrown into the fire on one of the *Anaglypha Trajani* reliefs are wax tablets.[38] These reliefs are particularly noteworthy because they coincide with the period Pliny describes as dominated by papyrus records.

However, securely identifying the material depicted is not so easy. The material held by the writer on the far left of the Ahenobarbus relief takes codex form. Rather than inset, the writing surface appears to be raised from the surround. The thickness of the books also seems greater than might be

32 Bowman & Thomas (1983) 44.
33 Suetonius, *Aug.* 85.2, *Claud.* 16; Athenaeus 9.407b (on the same problem in fifth-century Athens).
34 Vitruvius, *De Arch.* 7.10.
35 Vitruvius, *De Arch.* 6.4.1.
36 Bucher (1987) 25, 54n.75.
37 N° d'entrée LL 399 (n° usuel Ma 975), Collection & Louvre Palace. Pictured at Torelli (1982) I.4a and www.louvre.fr/en/oeuvre-notices/so-called-domitius-ahenobarbus-relief.
38 Housed in the Senate House, Rome. Pictured at Torelli (1982) IV.10.

imagined even for the bulky wax tablets.[39] However, if it is a codex pictured, it is a remarkably early example.[40] Further, the census records of the Republic are referred to as *tabulae*.[41] The carver may have been attempting to render a triptych-type tablet, with the books piled by his knee representing bundles of wax tablets. Thus the traditional interpretation of the relief can be upheld.

In the case of the Ahenobarbus relief, it is possible that the wax tablets represent a preliminary stage, with a full census list and its various derivatives compiled on a papyrus (or linen, see later) roll for storage. On the other hand, in the Trajanic relief the aim of the depicted exercise was to destroy tax records. Destroying only preliminary accounts or notes would not have achieved this.[42] Moreover, the Republican censors gave lists of taxpayers to the treasury.[43] Although the Trajanic reliefs belong to a later period, they suggest that these taxpayer records, derived from the census, were wax tablets. Both carvings indicate that wax tablets had an important archival function in ancient Rome, particularly in the context of the census. The records may have been bulky, but they nonetheless appear to have been the censors' material of choice.

Further, Meyer demonstrates that Romans considered wax a long-lasting material; for example, *imagines* of ancestors were sculpted from wax.[44] Records on wax tablets were also seen as more reliable. Cicero presents wills, contracts and accounts as having a greater legal status when on *tabulae*.[45] This significance probably originated from the treatment of *senatus consulta*. To be considered valid, *senatus consulta* had to be recorded and deposited in the *aerarium*.[46] Storing these tablets in the *aerarium* validated the advice because Saturn acted as guarantor for their contents.[47] The act of recording and depositing *senatus consulta* was more significant than the

---

39 Torelli (1982) 9 hedges his bets, referring to the writing materials as registers and books.

40 Torelli (1982) 15 dates the relief to 115 and the censorship of Cn. Domitius Ahenobarbus.

41 E.g., *tabulae* Livy 6.27.6, 29.37.7.

42 Posner (1972) 163 states that these tax records could not have been expected to last, but provides no argument. It is unlikely that tax records were expected to last a great many years.

43 Livy 29.37.12.

44 Meyer (2004) 35; Sallust, *Iug.* 4.6.

45 Gurd (2010) 85; cf. Cicero, *Ros. Com.* 2.6–7. Ulpian (*Dig.* 29.3.2.2) states that wills had legal status on whatever medium they are written. Cicero does not disagree, but does suggest that *tabulae* were considered (at least in the first century) more reliable.

46 Josephus, *AJ* 14.10.10 – Δόγμα συγκλήτου ἐκ τοῦ ταμιείου ἀντιγεγραμμένον ἐκ τῶν δέλτων τῶν δημοσίων τῶν ταμιευτικῶν Κοΐντω Ῥουτιλίω Κοΐντω Κορνηλίω, 'the opinion of the senate, copied from the treasury from the public tablets of the quaestors Quintus Rutilius and Quintus Cornelius' cf. Plutarch, *Cato Min.* 17.3; Suetonius, *Aug.* 94; Cicero, *Cat.* 1.2.4. 'Public tablets' is probably a direct translation of '*tabulae publicae*', Shrek (1969) 8 cf. Plutarch, *Cato Min* 17.3.

47 For temples as guarantors, see V:III.

material, much in the same way that storing census records was a significant part of the *lustrum* ceremony (see III:v). In Meyer's words, the deposited records and *tabulae* were the 'authoritative and final embodiments of the new reality they helped to create'.[48] Wax tablets held authority due to these associations. Cicero's trust in *tabulae* demonstrates that the line between the validity of material and of contents had blurred. The tablet itself became a guarantor of authenticity. Wax was considered a reliable, long-term archival material.

Wax tablet is the only form in which in situ archives survive. During the Pompeii excavations, business records belonging to an auctioneer named L. Caecilius Jucundus were discovered in his home.[49] These records were private rather than concerned with state governance, and do not reflect the scale of record keeping which accompanied the census. Nonetheless, the discovery is significant. Jucundus' tablets show signs of organisation. Although some of the archive seems to have been destroyed (perhaps in the earthquake of 62[50]), and attention was not paid to their organisation during the excavation, many of the records are marked on the edge.[51] Posner suggests that this was for quick identification and thus that the tablets had an order, forming a small but organised archive.[52] Jucundus' archival organisation supports a similar arrangement for *senatus consulta*. Coudry has pointed out that some inscribed laws include what appears to be an archival reference to the location of the relevant *senatus consultum*. The information is not relevant to the inscription's content but was copied by the engraver. This reference included the consular year, the month and a number specific to that *consultum*.[53] Josephus refers to a similar system, although with eponymous quaestors.[54] Together, this evidence points to the regular use of wax tablets to store information in an organised archive.

Whether military documents were stored on wax tablets can now be explored, although the lack of direct Mid-Republican evidence means that conclusions must remain hypothetical. Firstly, it is probable that quaestorial accounts were submitted to the treasury on wax tablets. Josephus refers to the public tablets of the quaestors with specific reference to *senatus consulta*, but his mention does not exclude other documents being amongst these 'public tablets'.[55] In 64, urban quaestor Cato the Younger was able to exercise his veto over his colleague to erase entries made by the latter on treasury tablets relating to monies owing, suggesting wax tablets over

48 Meyer (2004) 22.
49 *CIL* 4 Supp. 1.1–153.
50 Jongman (1988) 215.
51 E.g., *CIL* 4 Supp. 1.38, 100, 158. Cf. Andreau (1982) 14.
52 Posner (1972) 163.
53 Coudry (1994) 67–9; e.g., *CIL* 1.2^2.588.1.3.
54 Josephus, *AJ* 14.10.10.
55 Josephus, *AJ* 14.10.10.

another form and thus indicating that it was a regular form for public accounts.[56] Quaestorial campaign accounts acted (theoretically at least) as a guard against corruption on campaign (see III:vi, IV:v). As Cicero demonstrates, the wax tablets were considered more reliable (not to mention more durable on campaign).[57] Practical and ideological reasons suggest that quaestorial accounts were kept on wax tablets.

Secondly, the census. It has been established that recording the census involved wax tablets. However, it seems the derivative lists took the same form. Lists of those liable for military service were called the *tabulae iuniorum*; the use of *tabulae* points to wax tablets. These lists were organised by tribe; a full list for each tribe would have taken several tablets, especially if exemptions and military service were also marked. Alternatively, the *tabulae iuniorum* may have excluded anyone already exempt, shortening the list and speeding up the selection process at the *dilectus*. The bulk of these lists is not an argument against wax tablets. If the entire census was recorded on wax tablets, a derivative list would have been smaller and comparatively easier to use.

On the other hand, it is less clear whether the legion lists generated at the *dilectus* were written on wax tablets. The only source to explicitly mention the lists is Polybius. His Greek does not indicate a material, only that a list was generated.[58] As a copy went on campaign, it is possible that leaf-style tablets were used for ease of transportation. Alternatively, wax tablets were desirable due to the extra durability offered both in the field and (if necessary) for use at the next census. Either seems equally plausible. On balance, a wax-tablet list, at least in Rome, may have been more sensible, as the list would need to survive five years intact to be useful at the next census.

Linen

Pliny's final writing material is linen, which he describes an earlier alternative to papyrus. *Libri lintei* (linen books) are mentioned by several ancient writers, but modern scholars tend to consider linen principally the reserve of Etruscan religious writings.[59] The evidence does not support this conclusion. The surviving linen wrappings of the third-century Zagreb

56 Plutarch, *Cat. Min.* 16–8, 44; Pina Polo & Díaz Fernández (2019) 82.

57 It would thus follow that Verres' accounts (see IV:v) were also on wax tablets (Cicero, *Verr.* 2.1.14.36). Gurd argues that they in fact came from less reliable *commentarii* (Gurd (2010) 85, 90 using Cicero, *De Or.* 2.52). However, public storage on wax tablets would help to explain how Cicero got hold of them.

58 E.g., Polybius 6.21.1.

59 Livy 4.7.12, 4.13.7, 4.20.8, 4.23.2, 10.38.2; Fronto, *Ep.* 4.4.1; SHA, *Aurel.* 1.7; Ulpian, *Dig.* 28.1.22; Pallottino (1955) 153–4; Posner (1972) 164; Frier (1975) 88; Haines (1982) 175; contra Bucher (1987) 28–9.

mummy do contain what appears to be a religious Etruscan text.[60] However, in AD 144 Fronto reports that he discovered many religious ceremonies and linen texts at Anagnia.[61] Likewise, the formation of the Samnite Linen Legion, which involved religious formula from a linen book, was manifestly not Etrurian.[62]

Moreover, there is no need to limit linen's role to the purely religious. For example, the Linen Legion was formed using a religious linen book within a linen-covered compound, which gave the legion its name.[63] However, this religious role does not rule out an administrative one. As already seen (III:v), by the Middle Republic the *lustrum* ceremony of both the census and new commanders in the field was a religious event insolubly tied to written administration. Depositing the census documents was a ritual event; subsequently the documents could not be tampered with, indicating a sacred status. Further, the linen book used by the Samnites to form the Linen Legion had an explicit military connection. The priest claimed that the formula originated from ancient battle plans against the Etruscans. Thus, while linen had some level of sacred status in third-century Italy, this does not prevent a parallel administrative role. Indeed, the majority of Livy's references to *libri lintei* suggest a secular purpose.[64] The *libri lintei* may thus provide an earlier alternative to papyrus rolls, just as Pliny suggests.

The only known Roman linen records are those in which, according to Livy, Licinius Macer found a list of fifth-century magistrates which diverged from the established tradition.[65] If genuine, these *libri lintei* would demonstrate a very early form of administration recorded on linen. However, several modern scholars consider these *libri lintei* a hoax,[66] possibly perpetrated by Macer himself. Frier argues that Macer wrote his history in response to Claudius Quadrigarius' attack on the historical method of previous annalists.[67] Claudius apparently began his history after the Gallic Sack in an attempt to convey only what he considered reliable information,[68] effectively deleting the prominent fifth-century history of the Licinii. Such a motivation may have spurred Macer to 'discover' a lost record to

60   Van der Meer (2007).
61   Fronto, *Ep.* 4.4.1 – *praeterea multi libri lintei, quod ad sacra adtinet.*
62   Livy 10.38.5.
63   Livy 10.38.5.
64   Livy 4.7.12, 4.13.7, 4.20.8, 4.23.2.
65   Livy 4.7.12, 4.13.7, 4.23.2, 4.20.8 – *quis ea in re sit error, quod tam veteres annals quodque magistratuum libri, quos lintei in aede repositos Monetae Macer Licinius citat identidem auctores.* Cf. Dittman (1935) 288; Ogilvie (1958) 40; Meadows & Williams (2001) 29.
66   E.g., Mommsen (1859) 93–8; Gudeman (1894) 143; Klotz (1937) 217. For a fuller discussion of this issue and further bibliography, see *FRHist I* 324–6.
67   Frier (1975) 93–4 with Plutarch, *Num.* 1.2, contra *FRHist I* 324–5.
68   This accords with Livy's statement (6.1) that most of the city's records were destroyed in the fire which accompanied the Sack.

rehabilitate the history of his *gens* (although this is not the conclusion Frier reaches). Further, Richardson has highlighted suspicious patterns in the early magistracies held by several *gentes* of Roman historians.[69] There are strong arguments for disregarding Macer's *libri lintei*.

However, it is not the veracity of the content of Macer's discovery which is important here, but the use of linen. Livy mentions the *libri lintei* due to a disagreement between two of his sources, Tubero and Macer, over the office and identity of a certain year's magistrates. Livy questions the accuracy but not the existence of Macer's source. Significantly, as Ogilvie points out, Tubero does not question its existence either.[70] All three – Livy, Tubero and Macer – find the idea of administrative state records on linen entirely plausible. Indeed, if the rolls were a hoax, why would Macer have chosen an unprecedented material for his 'discovery'? None of this absolutely confirms that linen had an administrative role in Roman history, but it nonetheless points strongly towards it.

Further, it is worth considering the location of Macer's discovery: the Temple of Juno Moneta.[71] The temple was founded in 345.[72] The exact meaning and origin of 'Moneta' is unclear, with three etymological and historical origins suggested by the Romans themselves. Meadows and Williams are right to conclude that the true origin will probably never be known.[73] However, they argue convincingly that Moneta was strongly associated with the Greek Mnemosyne and memory, and should be translated as 'the Remembrancer'. Thus Juno Moneta was a credible source unlikely to be questioned.[74]

Moreover, temples were the natural home of records (see V:iii). This connection between temples, the sacred and documentation provides further support for linen as an archival material. If linen was readily associated with the religious (as the Linen Legion, the Zagreb mummy and Fronto all indicate), it suggests that it was an obvious choice for important state documents which required safeguarding. In particular, the census records, documents already loaded with religious associations, may have been recorded on linen (see later). Macer's *libri lintei* may have been stored in the Temple of Juno Moneta due to the religious connotations of linen as much as the Remembrancer nature of the temple itself.

---

69 Richardson (2014) 34.
70 Ogilvie (1958) 46.
71 Livy 4.7.12, 4.20.8.
72 Livy 7.28.4–6.
73 Meadows & Williams (2001) 33. Possible origins are: 1. *evocatio* from Veii (Livy 7.28.4–6, Plutarch, *Cam.* 36.9; Ovid, *Fasti* 6.183–90; Valerius Maximus 6.3.1a); 2. warning during an earthquake (Cicero, *Div.* 1.101); and 3. advice in a time of war (Suda s.v. Μονητα).
74 Meadows & Williams (2001) 33, 36–7; cf. Hardie (2007).

Linen was considered a sturdy material. The possibility of a fifth-century linen document surviving into the first century is not disputed by Macer's ancient detractors. The same attitude is visible regarding the linen of Cossus' *spolia opima* from 437. This inscribed linen breastplate was housed in the Temple of Jupiter Feretrius.[75] When rebuilding this temple, Augustus claimed to have seen the breastplate with its inscription still in situ.[76] Livy accepted Augustus as an authority. If Augustus' claim is true, this demonstrates that linen could survive nearly half a millennium intact and with its inscription still legible.[77] More significantly, it illustrates that Livy and his audience considered such a survival plausible. Even if linen could not survive intact for 500 years, the Romans believed in its longevity. Such a belief, coupled with its demonstrably sturdy nature, would have made linen an ideal archive material.

Were military records kept on linen? The census is the most obvious candidate. As established, wax tablets were used for collecting census data, but this does not rule out linen's use for the final definitive list which was carefully deposited, or for derivative lists. Following his defence of the early census figures, Frank has argued that the figures from before 225 were preserved in the same *libri lintei* consulted by Macer. Livy cites Macer and the *libri lintei* particularly for the few decades following the establishment of the censorship.[78] This may be a false correlation, as it is possible that Livy cited Macer and the *libri lintei* again in the lost books 11–20. Nonetheless, using the extant evidence, Frank is right to consider this correlation notable. It appears that the *libri lintei* may also have included (or been believed to include) census data alongside lists of magistrates. This perhaps ties the books more closely to the *Annales Maximi*, which included notable events of the year. Either way, this indicates that linen was considered a suitable material for recording such information. If a summary could be kept on linen, it is plausible that the more detailed breakdown required to mobilise this manpower outside a *tumultus* was recorded likewise.

---

75 Livy 4.19–20.

76 Livy 4.20.7.

77 This should perhaps be doubted, however. To survive into the 20s, the breastplate would have had to survive a devastating fire on the Capitol in the 70s. Further, if glue was used to laminate the armour's layers (cf. Aldrete (2015)), it may have had an effect on prolonging the life of the material. Alternatively, the only secure surviving example of linen armour was made using a twining method (Yale no. 1933.481 cf. Pfister & Bellinger (1945) 59, Taylor (2012) 64–71). This greave and the Zagreb mummy reveal that linen can be extremely long-lasting in the correct conditions.

78 Frank (1930) 316n.10.

Military documents

To conclude, Pliny's comments on writing materials help to establish the possible form of Mid-Republican military documentation. The lack of direct evidence means that these conclusions cannot be certain, but the weight of evidence is compelling. It is probable that leaf-style tablets were a favoured material for organising and regulating an army on the move. Their ease of production and small size would make them ideal for this task. As decisions regarding legion recruitment and deployment were made by the senate, such information was treated like any other *senatus consultum*, that is, written on a wax tablet and deposited in the *aerarium*. Financial accounts had a clear connection to the treasury and needed to be considered immutable, suggesting that a wax tablet was the ideal medium. Cicero's lack of trust in *commentarii* suggests not wax tablets but linen as the stored form of *commentarii*, although for those on campaign, leaf tablets may have been more convenient for a first draft. It is only with the census that a conclusion is more difficult. The repeated mention of wax tablets means that their presence cannot be ignored, but neither can the possibility that the full list could be written up on a more easily stored linen roll.

Why did papyrus take so long to become an archival material in Rome, allowing linen a perhaps-unexpected prominence? The answer is in part provided by Pliny's passage. It has already been commented that Pliny chose to use the term 'papyrus' only once in his discussion of writing material, preferring the term *charta* instead. The conclusions of this section suggest that the emphasis of *charta* should be 'writing surface' generally rather than papyrus in particular. Pliny emphasises the importance of papyrus as a writing surface because it had become the primary record form. The emphasis in the following phrase is that it was writing – that is, bureaucracy – that was important to running the empire, not papyrus in particular. In addition to supporting the notion of a relatively complex bureaucracy in the Middle Republic, Pliny's comments indicate the endurance of other writing surfaces, not requiring papyrus to be prominent until the first century AD. The often-noted conservatism of Rome may well explain the time taken for papyrus to reach ascendancy. Rome already had several serviceable writing surfaces; there was no pressing need to change a functioning system.

ii Record sizes

Before continuing to discuss storage locations, the potential size of records must be considered. An approximate volume of the records will help in attempting to identify possible archives. This is most easily achieved by examining wax *tabulae* and the legion list generated at the *dilectus*. Using a sample of Jucundus' tablets from Pompeii (with an average wax area of 82 × 107 mm and depth of 7 mm), Bucher has calculated that the average character density of the tablets is 19,226 characters per square metre, or approximately 167 characters per page.[79] Using a tablet of this size, the potential length of the legion list can be calculated. For an ordinary legion, 4500 names would be required. Latin names in the form 'C. Iulius Caesar' have approximately 16 characters on average, excluding spaces.[80] Thus, for a cramped list using all the available space on a tablet, 450 *tabulae* of 147 × 118 mm would be required. At 7 mm thick each, these would take up 3.15 m of shelf space. If four legions were in the field, and the legions were larger, the space required even for the military records of just one year quickly grows. Census records, which contained a much greater number of individuals and more information about them, would have taken up considerably more space.

However, the size of Jucundus' *tabulae* is not a limiting factor. Both the Ahenobarbus relief and the *Anaglypha Trajani* depict larger tablets.[81] The perspective of the carvings makes exact measurements impossible, but the relative size of nearby bodies allows an estimate. The *tabulae* appear to be approximately the length of a thigh and half as wide, that is, about 45 cm long and 22.5 cm wide. The open pair on the Ahenobarbus relief suggest that the border around the wax area is about the width of the standing figure's finger. A 2.5-cm border will be assumed. This gives a wax area of 42.5 × 20 cm. Following Bucher's calculations, the maximum number of characters per tablet is 1634 (to the nearest character).[82] For a cramped legion list using all the available space, 45 tablets would be needed. It is reasonable to assume that the thickness of these larger tablets would be greater than that of Jucundus' smaller ones, to prevent snapping: 10 mm will be allowed per tablet. Thus, 45 cm of shelf space would be required. The overall volume of the list on the larger-size tablet is also smaller, 0.045 cubic metres compared to 0.054 cubic metres. In terms of accessing the records in addition to storing them, the larger volumes would have been much more convenient, as they would use much less shelf space. These are of course estimates; any changes in the size of writing would have an impact on the number of *tabulae* required.

As it has been suggested that records on campaign were kept on leaf-style tablets, it is worth translating (as far as possible) these lists onto this material. The known tablets are approximately the size of postcards. Here a size of 10 × 15 cm will be assumed, although examples as wide as 25 cm are known. The tablets are approximately 1 mm thick. Size probably depended on the nature of available timber as well as the intended purpose. The same character density as wax *tabulae* will be used for the purposes of comparison. Thus the maximum number of characters per sheet is 288.[83] For a list of 4500 names

---

79 Bucher (1987) 25–6 using *CIL* 4 Supp. 1.1, 6, 7, 10, 17, 19, 21.
80 E.g., C. Iulius Caesar = 13, Ti. Sempronius Gracchus = 20, L. Aemilius Paulus = 15, Q. Lutatius Catulus = 16, Q. Fabius Maximus = 14, and so on.
81 See V:i, nn. 37–8.
82 = (0.425 × 0.2) × 19,226.
83 = (0.1 × 0.15) × 19226.

following the same conventions as before, 250 tablets would be required, taking up a shelf space of 25 cm and a volume of 0.004 cubic metres. This would be much more convenient for a travelling army disadvantaged by extra baggage. Even if each soldier were allocated half a sheet of the leaf tablet, the resultant records would only take up 0.034 cubic metres. Doing the same with the small wax tablets would give a volume of 0.273 cubic metres, eight times more.

These calculations all assume that *cognomina* were in regular use in the Mid-Republic, which is not necessarily the case. If it is instead assumed that names were given without a *cognomen* but with a filiation, significantly less space is required. Each name is now an average of approximately 11 characters.[84] Thus 297 small wax tablets, with a shelf space of 2.08 m, would be required for a legion of 4500, or 31 large tablets, with a shelf-space of 31 cm. For the leaf-style tablets with the maximum character density, 172 tablets would be required for the same legion. Overall, if *cognomina* were not used in lists, the quantity of material required was even less, and thus easier to store and transport.

Examining the size of records supports the conclusions reached before. The use of leaf-style tablets on campaign would be significantly more space-efficient than the wax equivalent. The use of larger wax tablets, as seen in the Ahenobarbus relief, for the census and its derivatives would have made storage indoors easier than use of a smaller version, both in absolute volume and in terms of accessibility. The lack of information about the nature of linen rolls means a similar analysis cannot be undertaken. Nonetheless, it can be said that information stored in this way would be more similar in size to leaf-style than to wax tablets.

iii Location of storage

The next element of Roman military administration to be examined is where these documents were kept. Identifying the buildings, locating them topographically, and examining their geographical as well as functional relationships all help to reveal the development of administration through the period. A lack of direct evidence concerning the Middle Republic means that there is a limit to how far this discussion can be taken. Nonetheless, this section examines possible storage locations, focusing on the complex of buildings at the northwestern end of the *forum Romanum* and the southeastern slope of the Capitol.

First, however, private storage must be addressed. Dionysius of Halicarnassus explicitly states that the section of the 393–2 census he saw was handed from father to son for preservation.[85] There is nothing to suggest that

---

84 E.g., C. Iulius C.f. = 9, Ti. Sempronius Ti.f. = 15, L. Aemilius L.f. = 11, Q. Lutatius Q.f. = 11, Q. Fabius Q.f. = 9, and so on.
85 Dion. Hal., *Ant. Rom.* 1.74.5.

Dionysius or his readers found this unusual. This raises several problems. Private storage is an obstacle to taking a new census based on the records of the previous *lustrum*. Any derivative lists could only be collated once every five years, failing to account for, in particular, movement in and out of the *iunior* age bracket (this problem is dealt with in detail later). Moreover, keeping census records privately contradicts the census record's ritual depositing, *lustrum condere* (III:v). Storing sanctified records which embodied the citizen body, and so in a sense Rome itself, in the house of a magistrate about to return to private life is incongruous. On the other hand, there is little reason to suspect Dionysius' discovery. A solution is required.

Rawson uses this passage to suggest that it was regular practice for censors to take documents home after their period of office as mementos of their achievement.[86] This interpretation is supported by Dionysius' language. His wording can be literally translated as 'from censorial memorials', ἐξ [...] τῶν [...] τιμητικῶν ὑπομνημάτων. This need not imply the entire census, or even large parts. Perhaps it involved selected highlights. Alternatively, Suolahti argues that censors kept a draft form of the census, particularly earlier in Rome's history.[87] Again, a full version is unlikely, if only due to the impracticality of storing it (see V:ii). For the census, at least, privately held records do not appear to have been the official state version.

Magisterial *commentarii*, however, were kept privately. Cicero encourages retiring magistrates to turn over their records to the censors,[88] implying that first-century magistrates did not do so habitually. The same appears to be true in the Middle Republic. Livy records Scipio Africanus destroying his account book in frustration when the senate questioned his integrity.[89] The passage is somewhat confusing, as it seems to conflate two separate bribery accusations against Lucius and Publius Scipio.[90] Africanus' case is nonetheless illuminating. In order to destroy his accounts, Publius sent Lucius to fetch them.[91] This suggests that the accounts were kept in the Scipio household rather than in an official archive such as the *aerarium*. There is no sense that this was unusual; it was their destruction which was shocking, not their storage location. Coupled with Cicero's appeal, it seems

---

86 Rawson (1985) 238–9.
87 Suolahti (1963) 33.
88 Cf. Mommsen (1893) 4 n.2; Cicero *Orat.* 46.156. This also implies that the censors had an archive of sorts in which to store these documents.
89 Livy 38.55.10–2 – *librumque rationis eius*. *Liber* perhaps indicates linen or papyrus as the material. Certainly it was something easily torn, ruling out wax tablets.
90 Cf., e.g., Luce (1977) 92–104; Jaeger (1997) 132–75. Livy's account suggests that neither Lucius nor his quaestor C. Furius Aculeo considered their actions illegal or immoral, especially as the state was the beneficiary before either was arraigned (Livy 38.55.5–7). On the other hand, Africanus' actions are more than a little suspicious.
91 Livy 38.55.10–1.

that the private storage of magisterial *commentarii* and commanders' financial accounts was not unusual.

Aerarium Saturni

The discussion can now turn to public storage. *Senatus consulta* were stored in the *aerarium Saturni,* or treasury of Saturn, on wax tablets; it has already been suggested that quaestorial financial accounts were also kept there. Quaestors functioned as treasury officers while on campaign, with their accounts (parallel to consular records) guarding against corruption (see IV:v). It follows that these accounts were then stored in the *aerarium,* both because it was effectively the quaestors' administrative hub and due to the importance of depositing documents in the treasury to ensure their validity. This would explain how Cicero was able to get hold of Verres' accounts, as it is unlikely Verres volunteered them from his private records.[92] Sutherland describes the *aerarium* as the heart of Rome's financial machinery.[93] The image of a machine is perhaps going too far, but Brunt is undoubtedly right to suggest that the quaestors running the treasury had an active role in advising the senate on financial matters rather than simply acting as custodians.[94] In order to organise and pay an army, some concept of state funds was required. Accounts under the care of the quaestors are a logical conclusion, even if they were more abbreviated in form than those previously proposed as having been kept on campaign (see IV:iv–v).

There are few ancient references to keeping documents besides *senatus consulta* in the treasury for validity, but temples were considered to provide secure and binding storage.[95] The importance of depositing census documents as a sanctification and safeguard of the citizen body has already been discussed (III:v). Other documents stored in temples are also recorded. The *libri lintei* in the Temple of Juno Moneta may be one such example (see V:i).[96] Julius Caesar's will was entrusted to the Vestal Virgins, presumably to keep it secure and prevent tampering.[97] That Octavian was able to manipulate Marcus Antonius' will as he did reinforces this presumption.[98] Beard suggests that lists of temple contents given by Pliny the Elder may be derived from a 'contents list' kept in each temple.[99] These do not date from the Middle

92 Cicero, *Verr.* 2.1.14.36, IV:v.
93 Sutherland (1945) 154.
94 Brunt (1966) 90, contra Millar (1964) 38. Cf. Polybius 23.14.5–6.
95 Cf. Culham (1989) 110.
96 Livy 4.7.12, 4.20.8.
97 Suetonius, *Iul.* 83.1.
98 Suetonius, *Ant.* 58.2–3; Dio 50.3.3–4.1. Whether Octavian invented the offending passages after forcing the Vestals to relinquish it is irrelevant.
99 Beard (1998) 93; e.g., Ceres: Pliny, *HN* 34.15, 35.24, 35.99; Concord: Pliny, *HN* 34.73, 77, 80, 89–90; Apollo Palatinus: Pliny, *HN* 36.13, 24–5, 32, 32.11.

Republic nor refer directly to documents, but together with the other examples they are indicative of a culture in which temples provided security.

It is also plausible that some quaestorial *commentarii* could have been kept in the *aerarium*.[100] Such records could have informed future incumbents, just as privately stored records assisted family. Reading *commentarii* would be a quick way to familiarise oneself with the requirements. As the *aerarium* was both an archive and the building most associated with the quaestors, it is the most obvious place for *commentarii* to be kept if they were not stored privately. Additionally, the need to hand over financial accounts may have made the passing of *commentarii* more natural than for other magistrates.

Moreover, the *aerarium Saturni* had a more direct link to military matters: the military standards were kept there when not in use.[101] This may point more to the importance of temples, or to Saturn in particular, as guardian forces than to a military link. Nonetheless, that the treasury had such a long-standing link with military paraphernalia suggests that over time, the storage of other military items in the same location would have seemed natural. *Senatus consulta* on recruitment, deployment and reinforcement were already deposited there as a matter of course, and it appears that financial accounts of campaigns were also stored there. Military spoils had long been deposited there, leading Culham to describe Saturn as a 'heavenly book-keeper'.[102] It is not a great leap to imagine that information such as legion lists could also have been archived in the *aerarium*. This cannot be conclusively demonstrated, but the *aerarium* remains a good candidate for the location of the storage of military documents in the Middle Republic.

Exactly where the *aerarium Saturni* was located is unclear, and much debated. It is strongly associated with the Temple of Saturn, which is located on the southeast slope of the Capitol. Since the erection of the building known as the 'Tabularium', the temple appears to be in the *forum* rather on the Capitol. However, Purcell highlights that Saturn is on the hill, as the *forum* began at the bottom of the slope.[103] At the beginning of the Middle Republic, the closest buildings in the *forum* were the senate house and the Temple of Castor and Pollux. Without the obstruction of the 'Tabularium', the connection with the Capitoline complex was much clearer (Figure V.1).[104] This in itself suggests a possible military connection, as it was on the Capitol that the *dilectus* took place (see I:iv), the auguries for consuls were taken and a campaign officially began.[105] The area around

100 Varro, *Ling.* 6.91.
101 Livy 3.69.8, 7.23.3.
102 Culham (1989) 111.
103 Purcell (1993) 132.
104 Richardson (1980) 53.

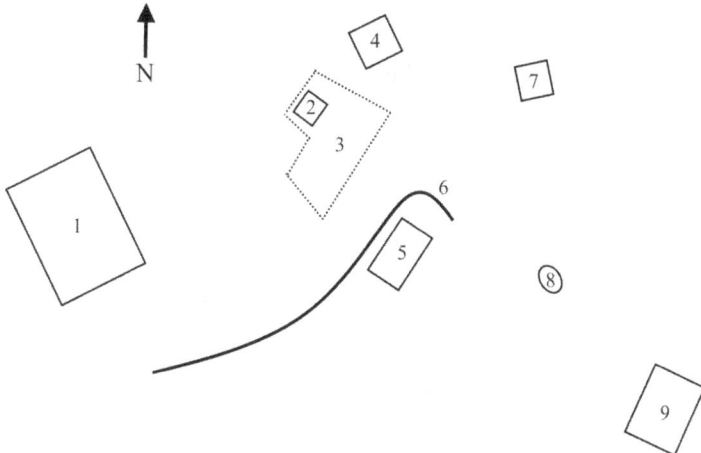

Figure V.1 Southeastern Capitol and northwestern *forum* in the third century.
1. Jupiter Optimus Maximus, 2. Mint(?), 3. 'Tabularium' (built 78),
4. Juno Moneta, 5. Saturn, 6. *Clivus Capitolinus*, 7. Senate House,
8. Lacus Curtius, 9. Castor and Pollux.

the temple, later built upon, probably made up the *area Saturni*, although its boundaries are unknown. The temple is one of the oldest in Rome; its exact date is disputed, but the foundation was probably sometime in the first decade of the fifth century, around the Republic's founding.[106]

Where the *aerarium* was in relation to the temple is disputed. One suggested location is underneath the stairs leading up to the podium.[107] This is supported by Asconius' comment that Pompey's chair was placed *ad aerarium* during Milo's trial.[108] That Ti. Gracchus sealed the door to the *aerarium* in 133 suggests that there was only one entrance.[109] Richardson takes this to mean that Pompey's chair was placed before the treasury's door, which he seems to imagine was in the side of the podium. However, there is no need to limit the translation of *ad aerarium* to 'before the door of the *aerarium*'. The placement of Pompey's chair could have been anywhere

---

105 Livy 45.39.
106 Tullius Hostilius or Tarquin Superbus: Macrobius, *Saturnalia* 1.8.1; Dion. Hal., *Ant. Rom.* 6.1.4; 501: Macrobius, *Saturnalia* 1.8.1; 498: Dion. Hal., *Ant. Rom.* 6.1.4; 497: Livy 2.21.2; Dion. Hal., *Ant. Rom.* 6.1.4. For a description of the temple, see Platner & Ashby (1929) 463–4; Coarelli (1999) 234–6.
107 Richardson (1992) 344, cf. Corbier (1974) 632. Under the podium of a temple was the location of the *aerarium* in Pompeii.
108 Asconius, *Mil.* 40 C cf. 41 C (*pro aerario*).
109 Plutarch, *Ti. Gracch* 10.6.

in the treasury's vicinity; its location is not narrowed down. Moreover, Asconius was writing after the temple's rebuilding in 42.[110] If the treasury, or part of it, was contained within the temple itself, the door may have been in a different place. Lugli reconstructs this space with a door in the side of the podium, but his drawing has been branded a 'preposterous reconstruction' with 'ugly design' and 'inadequate stairs'.[111] Design features aside, the stairs are problematic because they do not reach ground level across the hill's slope in front of the temple. The door is also truncated, making entrance to the space difficult. It does not appear to be the location of an archive.

There are also problems with the treasury being below the stairs in an early incarnation of the temple. Vaulting as an architectural feature was not used by the Romans until the second century.[112] Barring possible superficial repair work undertaken following the Gallic Sack,[113] there are not any notices of rebuilding Saturn until Munatius Plancus' work in 42. Thus, if there were a void under the steps of the temple, it would have been extremely cramped, poorly suited to the storage of coinage and metal, let alone documentation. It is unlikely that this was the site of a Mid-Republican archive.

Platner and Ashby suggest that only money, in whatever form, was kept in the temple *cella*, with documentation stored in an associated nearby building.[114] This is more plausible than storage under the steps. It is also likely that military standards were kept in the temple's *cella*. When Augustus placed the standards recovered from the Parthians in the Temple of Mars Ultor, they were deposited in the new temple's *cella*.[115] A god's guardianship was desirable for standards. As the objects themselves were symbolic, storage in the temple proper was necessary. In the case of documents, it appears that the act of depositing was more significant than the documents themselves (see V:i). Practical requirements insisted that documents be stored outside the temple proper. The storage duration is unknown, but there is no evidence of anything equivalent to the seven-year rule used today. Indeed, that Valerius Antias was able to consult senatorial records to compose his history indicates that they could potentially be stored for centuries (see III:ii). Thus, in searching for the treasury archive's location, a building of reasonable size close to the Temple of Saturn must be sought.

110 Suetonius, *Aug.* 29.5.
111 Lugli (1947) 35 fig. 4; Richardson (1980) 57.
112 Cf. Richardson (1980) 56.
113 Cf. Roberts (1918) 58.
114 Platner and Ashby (1929) 464.
115 Suetonius, *Aug.* 29.2.

The building known as the Southwest Building (SWB) is a promising candidate. This building, no longer surviving, was located across the *clivus Capitolinus* from the Temple of Saturn to the west, in the area on which the *Porticus Deorum Consentium* was later built (Figure V.2). The SWB pre-dated the 'Tabularium', which was built in 78, leaving a mark on the southern end of the façade.[116] Its exact construction date is unknown, but the late third or second century is likely.[117] The building's location, close to the Temple of Saturn and up the slope rather than in the *forum*, points to an association with the temple. Had the SWB been located down the slope in the *forum*, it could not be considered connected to the *aerarium*, as it would have existed in a different space. The period of its construction reflects the imperial expansion begun during the Middle Republic, a period which would necessarily have generated more paperwork from the senate and armies. Thus the SWB, although nothing more is known about it, fits in with the Republic's broader development as a building intended for document storage.

It is possible that the remains of a building excavated in a void of the foundations of the 'Tabularium' behind the Temple of Veiovis are also connected with the SWB. These consist of ashlar walls and mosaic floors, both apparently badly damaged in antiquity.[118] However, the author leans towards Tucci's argument that this second unknown building should be identified as the mint established c. 269 (Figure V.2).[119] It was perhaps subsumed into the 'Tabularium' following the fire on the Capitol in 83. There is an obvious connection between the mint and the treasury, and indeed by extension the military. This connection probably helped the eventual development of a complex which included the Temples of Saturn and Juno Moneta, the mint, the confusing 'Tabularium' and possibly the *atrium Libertatis*. However, if the void building functioned as a mint, it was not the storehouse of military documents and need not be further discussed.

The 'Tabularium' has been frequently mentioned in connection with the *aerarium* and deserves some comment. The building itself was constructed in 78, outside the scope of this study. However, as mentioned, what preceded it may be significant. Only two floors of the original building survive, revealing a confusing complex of corridors and tunnels. These are best laid out in the drawings of Purcell and Tucci.[120] Of particular interest is the lowest corridor, which connects to a series of rooms in the building's north

---

116 Tucci (2005) 9.
117 Tucci (2005) 21. It may be part of the rebuilding works done in the area in 174, although it is not explicitly mentioned (Livy 41.27.7).
118 Tucci (2005) 21; Sommella Mura (1981) 128–9, fig. 4.
119 Tucci (2005) 10. The proximity to Juno Moneta and possible role of the lowest, separate, level of the 'Tabularium' are the strongest factors; see later.
120 Purcell (1993) 136; Tucci (2005) 6. The author is not convinced by Coarelli's reconstruction of the form and function of the 'Tabularium', Coarelli (2010).

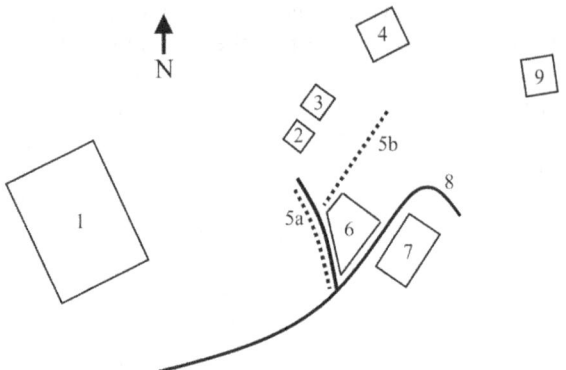

Figure V.2 Southeastern Capitol c. the first quarter of the second century, 1.
1. Jupiter Optimus Maximus, 2. Veiovis, 3. Mint(?), 4. Juno Moneta,
5a, 5b. *Porticus Saturni*(?) 6. Southwest Building, 7. Saturn, 8. *Clivus
Capitolinus*, 9. Senate House.

end and exited on the *clivus Capitolinus* to the south, just above the SWB.
This corridor and the rooms are apparently otherwise unconnected to the
rest of the 'Tabularium' above. They may well be associated with the mint
and connecting it to the *aerarium*.[121] For this study, the possible origin of
this corridor is significant.

In 174 a portico was added to the Temple of Saturn.[122] This is traditionally thought to have run from the temple to the *area Capitolina* along
the *clivus Capitolinus*.[123] Such an interpretation supports the idea that legion lists were kept in the *aerarium Saturni*, as they could be easily carried
to the SWB for storage following their production in the *area Capitolina*
(although the course of the path already allowed for this). However,
Richardson has suggested that the portico ran instead towards the senate
house along the slope, at roughly the later location of the 'Tabularium'
(Figure V.2).[124] As Richardson notes, this reconstruction requires a re-
interpretation of Livy, not a correction. Livy's description of the censors'
work is unclear, suggesting that he was describing something that he had
not seen and presumably no longer existed. If the portico, or part of it, ran
along the line suggested by Richardson, it may have been the first incarnation of a path leading from the mint to the treasury. This is indicative
of the growing association of buildings in this area. When the 'Tabularium'
was constructed, the route connecting these two places was maintained and

---

121 Coarelli (2010) 121–3.
122 Livy 41.27.7.
123 Platner & Ashby (1929) 463.
124 Richardson (1980) 62.

incorporated. This remains hypothetical, but points to a sense of connected and organised central administration reflected in architecture. The Temple of Saturn maintains its sense of guardianship, but this is extended through the developing complex.

Modern scholars often assume that the census documents were stored in the *aerarium* during the *lustrum*.[125] However, this location is mentioned in a specific context. It is only the list of *aerarii* (taxpayers) which is deposited in the treasury.[126] Prior to the cessation of citizen tax, this was the majority of those making declarations at the census. There is no reason to believe that this list was identical to the full census document, as only those *sui iuris* were liable for taxation. A separate taxpayer list would have been useful for tax collectors even when this was a more significant part of the population. Only information concerning names, location and tax to be paid was needed, not the extra information concerning family members and military service. This does not rule out storing the census documents in the *aerarium*, but it does not provide direct evidence for it. At the very least, a separate taxpayer list was also given alongside the full census. It is this act which Livy describes; the censor is not performing the *lustrum*.[127] Other possible locations for the census records must be examined.

Aedes Nympharum

It is better to focus on buildings more explicitly connected with storing the census: the *aedes Nympharum* and the *atrium Libertatis*. First, the *aedes Nympharum*. Cicero thrice mentions the temple as the home of the census records in his attacks on Clodius, who razed the temple in 57.[128] Nicolet argues that Cicero is referring not to the census records but to grain-distribution documentation.[129] Cicero states that Clodius destroyed the temple in order to destroy records kept there, '*ut memoriam publicam recensionis tabulis publicis impressam exstingueret*'.[130] The *De Haruspicum Responso* has a more roundabout reference to the event, but the *Pro Caelio* refers directly to the *censum populi Romani* and the *memoriam publicam*

125 E.g., Suolahti (1963) 33. Implied: Mommsen (1894) 245–9; Millar (1964) 34–6; Coudry (1994) 65; Coarelli (1999) 234. Contra Rawson (1985) 239; Culham (1989) 104; Nicolet (2000) 201; Meyer (2004) 29. The author is not aware of any scholar who makes an argument to support their assumptions.

126 Livy 29.37.12; Mommsen (1894) 249. There is no need to translate this as 'debtors' here, contra Millar (1964) 36.

127 Livy 29.37.12. The great antagonism between the censors in this year, with each trying to degrade the other into the *aerarii*, may explain Livy's explicit mention of this act, but it was not the *lustrum*.

128 Cicero, *Mil.* 73, *Har. Resp.* 57, *Cael.* 78.

129 Nicolet (1980) 64.

130 Cicero, *Mil.* 73. Nicolet considers these records to be grain-distribution rolls because the term *recensionis* is a rare one in Cicero and is used by Suetonius in this context.

being the object of Clodius' attack.[131] The combination of these passages makes it clear that Cicero is discussing the destruction of census records, not any other list, kept in the *aedes Nympharum*. This demonstrates that the census records were kept here in the first century, but further examination is required before the same can be concluded for the Middle Republic.

Beyond Cicero, the temple is unknown in the extant sources, making its Mid-Republican role difficult to ascertain. The temple was located somewhere on the *campus Martius*, probably close to the *Villa Publica*, where the census took place. The *aedes Nympharum* is usually associated with the temple on the Via delle Botteghe Oscure, diagonally across the crossroads from the *Largo Argentina*.[132] This temple has three building phases: a second-century foundation, a rebuilding towards the end of the Republic and finally Flavian work. The Late Republican work matches Cicero's reports of the temple's destruction. This does not confirm the temple's identification, but with other temples in the vicinity more securely identified, on balance it seems likely. Ziolkowski has argued that the foundation of the *aedes Nympharum* should be dated to 179–67; the foundation was recorded by Livy but has been lost in the lacunae of the last extant books.[133] Ziolkowski does allow that it could also have been built at any time following 179 down to the first century. This fits with the founding of the temple on the Via delle Botteghe Oscure, giving the earliest date possible as 179. Perhaps it should be attributed to the work of the 174 censors.[134]

At what point did the *aedes Nympharum* become the home of the census records? No modern scholar of whom the author is aware has offered an answer to this question. If the temple was intended from its inception as the storage location for census records, the answer is the same as its second-century dedication. However, there are several problems with this theory. Census record storage required a lot of room (see V:ii). While the *cella* of the *aedes Nympharum* may have been a symbolically safe place for storage, it is unlikely that it was practically suited to storing the records of cumulative *lustra*. Ideologically, it is also an odd choice to keep the written embodiment of the Roman people outside the *pomerium*, the city's sacred boundary. The census, with its origins as a military review, took place on the *campus Martius* to prevent arms being carried across the *pomerium*, but it is unlikely that census documents would be ritually deposited in this ideologically more vulnerable area. An alternative location seems more probable.

---

131 Cicero, *Cael.* 78 – [Clodius] *qui dedes sacros, qui censum populi Romani, qui memoriam publicam suis manibus incendit.*
132 Ziolkowski (1992) 121; Coarelli (1997) 222–3.
133 Ziolkowski (1992) 120, 311.
134 Livy 41.27.7.

The solution to this problem is possibly offered by several modern scholars, although they do not discuss their assumptions further. Suolahti suggests that records kept in the *aedes Nympharum* were those of the ongoing census before their transfer to permanent storage following the *lustrum*.[135] Similarly, Ziolkowski refers to 'some' records being kept in the temple.[136] That there was not a census in 57 when Clodius destroyed the temple is not problematic. The censors of 61–0 never performed a *lustrum*; the beginnings of their work may have still been in the temple in 57 for Clodius to destroy. Alternatively, Cicero may simply have been exaggerating the extent of Clodius' destruction, something which was not beyond him. Thus it appears that the *aedes Nympharum* was not the permanent storage location for the census records at any point in the Republic. Another location is required.

Atrium Libertatis

The *atrium Libertatis* is the other location repeatedly mentioned in the context of the census. This was where the review of those *equo publico* occurred.[137] The building is often considered a 'headquarters' for the censors, housing their offices.[138] This appears to be confirmed by the censors' actions in 169, shutting up the *atrium* with themselves inside over a dispute.[139] They locked away the public tables and closed the record room.[140] This action halted the census taking, as Livy states. This suggests that there were census records kept in the *atrium Libertatis* in a room designated for that purpose.

On the other hand, that the censors also sent away public slaves suggests that the records kept in the building may only have been those for the ongoing census, not ones permanently stored after the *lustrum*. This creates a conflict with the proposed role of the *aedes Nympharum*. Having two temporary census storage locations seems unlikely. However, there is a solution. Firstly, the foundation date of the *aedes Nympharum* is uncertain; it may not have been built by 169. Secondly, Livy's description indicates that the *atrium* was not just an archive. The presence of public slaves may suggest that census work was done in the building, perhaps the production of the final list or derivatives such as the *tabulae iuniorum*. Alternatively, these slaves may have been archivists helping the censors find and order necessary documentation from previous *lustra*.[141] Either way, the presence

---

135 Suolahti (1963) 33.
136 Ziolkowski (1992) 120.
137 Plutarch, *Pomp.* 22.4–6; Livy 29.37.8.
138 Platner & Ashby (1929) 56; Richardson (1992) 41; Coarelli (1993) 133; Purcell (1993) 143; Dix (1994) 283.
139 Livy 43.16.13.
140 Livy 43.16.13 – *obsignatis tabellis publicis clausoque tabulario.*
141 For more on permanent bureaucratic staff see VI:ii, iii.

of the slaves and a *tabularium* (a record room, not to be confused with the 'Tabularium') points to a more significant role for the *atrium*. The use of the *aedes Nympharum* by the first century may have been as a holding location until the records were needed by the *atrium*'s workers. The extra space may have been required following the massive increase in citizen numbers after the Social War.[142] This cannot be proven, but seems the most plausible explanation. The case for the *atrium Libertatis* as the census records' home cannot be dismissed.

The very name of the building, the 'Hall of Liberty', also suggests that it was the permanent storage location for census records. Purcell suggests that the *atrium* was the location in which non-Romans and ex-slaves were admitted to the citizenship.[143] Certainly it was the location where lists of freedmen's tribal allocations were posted in 167.[144] As the censors were responsible for allocating new citizens, their office is not a surprising location for this. However, that freedmen received citizenship on manumission suggests that the *atrium Libertatis* was the location where manumission was registered. Moreover, loss of Roman citizenship came in the form of exile and often (officially, at least) sale into slavery.[145] The census was a list of Roman citizens and thus, in Roman eyes, those who were truly free. A building dedicated to Liberty seems an excellent place to store the physical embodiment of the division between free and less free/slaves.

Further, the *atrium Libertatis*' physical history again suggests that it was related to document storage. Asinius Pollio restored it in 39–28, establishing Rome's first public library there.[146] Caesar had planned to provide public libraries containing, among other thing, a digest of legal codes.[147] Boyd considers it probable that this was included in the *atrium*'s library.[148] Ovid reveals that it held poetry, suggesting that the library contained a variety of texts.[149] The choice to establish a library in the *atrium Libertatis* rather than at a different location suggests an existing association with archival holdings. Coarelli thinks it probable that Pollio's library was the censorial archive's descendant.[150] Livy demonstrates that the *atrium* contained a *tabularium*. Moreover, the building was expanded by the 194 censors, when work on the *Villa Publica*, where census declarations were given, also commenced.[151] These expansions point to an increased censorial

---

142 [Livy], *Periochae* 98.
143 Purcell (1993) 143.
144 Livy 45.15.1–5.
145 *XII Tables* 3.5; Augustinus, *Civ. Dei* 21.11.
146 Suetonius, *Aug.* 29.5.
147 Suetonius, *Iul.* 44.1–3.
148 Boyd (1915) 31.
149 Ovid, *Tristia* 3.1.59–72.
150 Coarelli (1993) 134.
151 Livy 34.44.5.

workload and an attendant increase in paperwork production. Indeed, if space became so tight by the mid-first century that the *aedes Nympharum* was required to store some records, an association of the *atrium* with archives is probable.[152] Thus, the *atrium Libertatis* was probably the permanent storage location for the census records after the *lustrum*.

However, the whereabouts of the *atrium Libertatis* is unknown. Ancient writers had no need to describe the location of a familiar building; no modern scholar has proposed an undisputed site.[153] No comment can be made on its capacity, but the cumulative literary references suggest that there was room to store the census documents. Nevertheless, it is worth discussing one of the proposed sites. Purcell has suggested that the *atrium* composed the floors of the 'Tabularium' above the separated lower corridor, and was the building which stood on the site prior to the fire (Figure V.3).[154] His argument is intriguing, although he himself admits that it cannot be proved one way or the other. Interestingly, Boyd suggests that the library on the Capitol described by Orosius (in relation to Commodus' reign) was an old and venerable one.[155] He does not argue that this library is the 'Tabularium', but if Purcell's identification of the 'Tabularium' with the *atrium Libertatis* is correct, it would fit Orosius' description. Rome's oldest library, housed in a building associated for centuries with record keeping, matches Orosius' description; its destruction would have been noteworthy.

Towards an administrative complex

The identification of the *atrium Libertatis* with the site of the 'Tabularium' has a more interesting significance for this study. As Figure V.3 demonstrates, placing the censor's record office between the possible site of the mint and the Temple of Saturn creates a complex of buildings on the Capitol's southeastern slope all closely associated with administration. Although it is going too far to suggest that this was a planned administrative centre, the development of these buildings close to one another is unlikely to be purely coincidence. Juno Moneta and her mint were natural allies of Saturn and the treasury; their physical closeness reflects this, as does the possible location of the *porticus Saturni* joining the two across the slope. The SWB was a necessary extension caused by a development in Rome's bureaucracy. The *atrium Libertatis*' possible location on the site of the 'Tabularium' forms a centre point, just as it was the centre point of

---

152 The destruction of the *aedes Nympharum* in 57 may have helped spur the rebuilding of the *atrium Libertatis* on a larger scale.

153 E.g., Boyd (1915) 3–4; Purcell (1993); Coarelli (2010).

154 Purcell (1993). Contra esp. Coarelli (2010) 107–32. The author finds Purcell's argument more convincing than Coarelli's reconstruction of a building topped by three temples, but neither is definitive.

155 Boyd (1915) 19–20; Orosius 7.16.3.

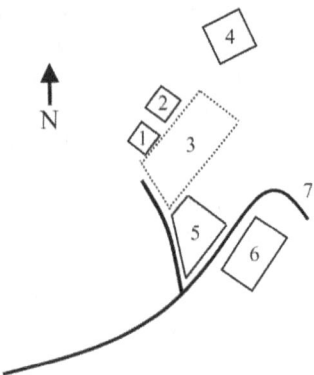

Figure V.3 Southeastern Capitol c. the first quarter of the second century, 2.
1. Veiovis, 2. Mint(?), 3. *Atrium Libertatis*(?), 4. Juno Moneta,
5. Southwest Building, 6. Saturn, 7. *Clivus Capitolinus*.

Roman citizenship. Censors could easily deposit the list of *aerarii* in the *aerarium*, along with any of the derivative lists stored there. Interaction between the two institutions would not have been limited by physical barriers.

The complex's central nature is also significant from the perspective of military administration. All the buildings in the complex had a link to the military: the temple and treasury of Saturn as the home of the military standards and other spoils; the mint by utilising the spoils; and the *atrium Libertatis* as the home of the census documents which formed Rome's manpower record. Documents from the beginning to the end of military campaigns and military careers were all stored in this area. Both the *atrium Libertatis* and the *aerarium Saturni* provided easy access to the *area Capitolina*, the location of the *dilectus* and official starting point of any campaign. Whichever building they were stored in, *tabulae iuniorum* could be quickly taken to the *dilectus*, and the legion lists generated there swiftly stored securely. Quaestorial financial accounts, some *commentarii* and the *senatus consulta* recording the decisions on deployment, collectively recording Rome's military activities, were all held in the *aerarium*. It is wrong to state that Rome had a military archive, but the collective work of these institutions did generate a documentary record of Rome's military activity. The buildings which contained it formed a complex, but one due to organic growth rather than deliberate planning. Despite this, it must be concluded that Mid-Republican Rome did have a bureaucratic military organisation.

The building of locations in which military documents were stored also reveals more about the development of military administration. It is striking that all the rebuilding, expansions, extensions and new foundations discussed in this section occurred in the first quarter of the second century.

The censorships of 194 and 174 stand out, but buildings less specifically dated, like the SWB, also belong to this period. This period of building could have had several causes; the influx of wealth which accompanied increased operations in the east and in Spain cannot be overlooked. Despite this, the choice to build the types of buildings specifically related to military administration (rather than, for example, temples) points to a specific trigger. The expansion of archival buildings fits with the increased levels of military documentation generated as a result of developments in the Hannibalic War. The regular keeping of legion lists in Rome following 204 would have created extra tablets to be stored, especially as Rome now had more legions in the field each year (see II:ii). The development of the complex on the Capitol's slope is in keeping with the emerging picture of developing military administration during the Middle Republic.

6 Record producers and record keepers

The final area to be addressed in this study is who was responsible for producing and storing the proposed documents. Only by examining this final human element can the picture of military administration in the Middle Republic be completed. The very existence of the records demonstrates that they could be produced. This chapter examines the human element in three sections. Firstly, literacy levels in the Middle Republic will be discussed, demonstrating that there were sufficient individuals capable of reading and writing at the level required by the proposed administration. Secondly, the individuals responsible for creating and caring for these documents both in the field and in Rome will be considered, examining the men who performed clerical roles for generals and quaestors. The section argues that military tribunes and in particular *scribae* generated the majority of the paperwork associated with the legions, including the legion lists and the pay and performance records based on them. Finally, the third section examines who was responsible for creating the census documents and their derivatives, especially the *tabulae iuniorum*. Combining the conclusions from these discussions demonstrates that Rome had the clerical capacity to generate and maintain military administration at the levels required by its growing influence. Rome developed mechanisms in the Middle Republic which would become those known from the Principate.

i Literacy in the Middle Republic

Plenty of work has been done on ancient literacy levels, reaching conclusions from relatively widespread literacy to levels as low as 1%.[1] This discussion has little to add to such studies as a whole, but it is necessary to establish the place of this monograph on the continuum. This section examines both the literacy level present in the Mid-Republican army and that required to operate its administrative functions. It demonstrates that while

---

1 E.g., Best (1966); Harris (1989); Hanson (1991) 159–60; Bowman (1991) 126, (1994) 111–2; Woolf (2009) 46.

high-level literacy may have been relatively minimal in Mid-Republican Rome, the army needed few to be fully literate; semiliteracy was sufficient.

First, it is worth examining the definitions of literacy in Harris' *Ancient Literacy*. Harris defines 'full literacy' as the ability to read and write complicated text without difficulty. 'Semi-literacy' is a broader category, encompassing anyone who has some ability in either reading or writing, for example, simple reading but not writing, or at the very lowest, only the ability to write their name.[2] As will be seen, the majority serving in the army fell into this semiliterate group. Harris uses two more terms to refer to types of literacy: 'scribal literacy' and 'craftsman's literacy'. Scribal literacy is defined as that of a specialised group literate for administrative purposes, almost a civil service, for want of a better term. Craftsman's literacy is when the majority of craftsmen are literate, but women, unskilled labourers and peasants are not.[3] Leaving aside the problem of using the term 'peasant' in the ancient world, the majority of Rome's population were farmers. The manpower recruitment problems caused by men's commitments as farmers and providers for families during the Hannibalic War demonstrates that this was still true in the late third century.[4] Thus, by Harris' definitions, the majority of the population was illiterate, with only a few craftsmen, scribes and the highest class (Harris does not define this, but presumably he means the equestrian centuries and senators) literate.

Such a conclusion – that only the highest in society and dedicated administrators were literate – is not necessarily invalid, but it is better to examine the evidence for literacy than to rely on one-size-fits-all models. For example, holding the military tribunate was based, at least in theory, on serving for at least five campaign years.[5] A capable man could, admittedly again in theory, rise through the ranks to hold the position. The office required literacy (see later), and Harris considers it unthinkable that a military tribune would be illiterate.[6] Either there were more conditions for the office than Polybius reveals (it is plausible that a higher class, and thus better education, were expected) or complete illiteracy was not as endemic as Harris suggests. Whichever was the case, the question of literacy requires more investigation based on the evidence.

The most-discussed evidence concerning literacy in the army is Polybius' description of passing watchwords and checking guard posts in the military digression of book 6 of his *Histories*.[7] Here Polybius describes how watchwords were passed from the tribune and onwards through the army

2 Harris (1989) 5.
3 Harris (1989) 7–8.
4 See I:v.
5 Polybius 6.19.1.
6 Harris (1989) 158–9.
7 Polybius 6.34.7–12, 6.35.5–36.8.

on small wooden tablets, *tesserae*. The tribune knew if the watchword had not made it through the camp by the failure of a *tessera* to return. Likewise, those inspecting the watch were issued written orders of when and where to inspect, and collected a *tessera* from the sentries. The tribune was able to tell if a watchman had been absent or asleep, and who he was, from both missing *tesserae* and the marks made on those which were returned to him. These measures demonstrate that written information formed the basis of security procedures, informing the tribunes about who had and had not received important information or performed crucial functions. The system is a relatively simple one, suggesting that it was introduced early in order to combat camp security problems. Best highlights that Pliny the Elder attributed the origin of the written watchword to the Trojan War, indicating that the method was so old that its origins were lost.[8] If writing was thus an integral part of military procedure, it required some literacy from many members of the army. More significantly, it implies an expectation of literacy from a relatively early period.

Best argues that these passages of Polybius, and Livian examples of orders on *tesserae*,[9] demonstrate a high level of literacy throughout the army. Every man needed to be able to read in order for passing written information to be successful.[10] However, as Harris has highlighted, Polybius' description does not involve all the legionaries.[11] Rather, watchwords were passed from centurion to centurion along to the tribune, with no indication that the watchword was handed round the entire century. Indeed, that the watchword was collected at sunset to be returned to the tribune by full dark suggests that there was not time for it to be passed around to each individual.[12] Instead, the best sense to be made of Polybius' description is that the centurions passed the watchword to their century orally once the tablet had moved on, requiring only the centurions and the tribune to be literate. Moreover, while the centurions made an identifying mark on the *tesserae*, it is unlikely this was of any great length – perhaps their name or their century's designation. The ability to read a short sentence and make an identifying mark falls under the category of semiliteracy.

However, this system does not confirm that all centurions were literate. It is possible that it could function so long as someone within the century could read and assist the centurion, a possibility made more plausible by the requirement that there be witnesses to the transfer of the watchword tablet. Further, Polybius only states that the centurion had to make his

---

8 Pliny, *HN* 7.56.202; Best (1966) 125.
9 Livy 7.35.1; 9.32.4; 27.46.1; 28.14.7.
10 Best (1966) 122–7.
11 Harris (1989) 167.
12 Polybius 6.34.8, 10.

mark, ἐπιγραφη.[13] Despite this, the sense of the passage remains that the centurions were semiliterate. Military tribunes were able to identify where in the camp a watchword was delayed from the ἐπιγραφαι, suggesting that the marks were distinguishable and attributable. Additionally, the precaution of a written watchword indicates an active desire to avoid miscommunications. The witnesses may also have helped avoid error, but this in itself again suggests some literacy. For the written version to have had value, semiliteracy was required of those involved in the process. Thus the watchword demonstrates that a high level of literacy was not required among the legions even at the rank of centurion, but nonetheless that semiliteracy was required of centurions.

On the other hand, the description of inspecting the watch suggests that there may have been literacy among all ranks. *Equites* inspected the watch, and the *veles* watchman provided a chit to prove he was awake and present.[14] This spans the army's entire wealth distribution. However, the *equites* may well have been the designated inspectors precisely because their richer background provided a better education, allowing them to read the more complicated written orders issued by the tribune.[15] This seems to be confirmed by Polybius' assumption that the chosen men would be able to read. Alternatively, as the inspectors appear to have been selected not by lot but individually,[16] it is possible that literate men were sought out. Nevertheless, the choice to regularly use the richer cavalry to inspect the watch indicates an expectation that a reasonable proportion had the requisite literacy skills.

The issue of the *velites* is less certain. As the youngest and poorest members of the army, it might be expected that they were also the least literate. The *tesserae* they gave to the watch inspector were supplied by the military tribune earlier in the evening, already marked with the watch and post.[17] As the guards were collected for their posts by an *optio*, it is unclear whether they read the slip themselves (although this does suggest that *optiones* as well as centurions were semiliterate). It does not demonstrate whether *velites* were expected to be, or were, literate. That the *equites* received written orders each night suggests that the inspection order varied, preventing the sentries from anticipating it. The regular nature of keeping watch meant that special written orders were not required for the *velites*, but it does not necessarily follow that the *velites* were incapable of reading them. Those *velites* who might be promoted to another battle line and those

---

13 Polybius 6.34.12.
14 Polybius 6.35.5–6, 8.
15 Although, as Polly Low (pers. comm.) points out, a system requiring writing could have been developed because equestrians were fully literate. It is probable that each factor reinforced the other as the system developed.
16 Polybius 6.35.8.
17 Polybius 6.35.6.

from higher-status backgrounds may well have been semi- or even fully literate. The use of *equites* for the inspection indicates an expectation of higher literacy among the richer cavalry, but complete illiteracy amongst even the poorest and lightest-armed cannot be assumed.[18]

Again, orders passed by *tesserae* suggest an expectation of some literacy among the soldiery.[19] However, it seems that here, as with the watchwords, only officers were required to read them. Best states that in 207 the consul Livius addressed his written orders to the entire army, although he admits it is unclear whether all could actually read them.[20] The wording is not as clear as Best suggests: Livy simply states that orders were sent through the camp, *tessera per castra*, probably in the same manner as the watchwords. It was much more convenient to pass the orders to the centuries orally than to wait for every man in the legion to read them, even if everyone was fully literate. Moreover, in all these cases the aim was to avoid using trumpets, which could alert the enemy to Roman intentions. Best goes too far in suggesting that the first recorded example of written orders, from the surrounded military tribune Decius in the First Samnite War, was to avoid the encircling enemy overhearing verbal commands.[21] If the enemy had been close enough to hear, the tribune's men would already have been overrun. It is implausible that a legion could operate in complete silence, but avoiding using trumpets – as in the case of Spain in 185[22] – would prevent a clear signal to the enemy of location and intention. Again, it appears that, among the foot, no one below the rank of centurion or *optio* was expected to be literate.

Further, it is probable that semiliteracy was all that was required of centurions. Orders which were ordinarily given by bugle call are unlikely to have been very complicated. Indeed, relying on individual subunits to simultaneously follow a complicated series of orders without further prompting, while not impossible, would have risked a breakdown in the legion's effectiveness. Thus, written orders in place of trumpets probably consisted of a few simple terms with which a semiliterate individual could quickly become familiar. The system assumes only a basic literacy.

More recent modern scholarship has argued for a higher level of literacy in more strata of Roman society than previously imagined, especially prior

18 Inscribed lead sling shot may be evidence of literacy lower in the ranks, especially if Greep (1987) 190 is correct in suggesting that shot was manufactured on campaign. This appears to have been a Republican phenomenon in Rome. However, inscriptions were generally cast on shot. While it suggests some literacy, it does not necessarily further demonstrate widespread literacy. For further bibliography see Greep (1987) 190.

19 Livy 7.35.1; 9.32.4; 27.46.1; 28.14.7.

20 Livy 27.46.1; Best (1966) 124. Best identifies the wrong consul; Claudius Nero was not yet at the camp.

21 Livy 7.35.1; Best (1966) 123.

22 Livy 39.30.4.

to the second century. Cornell and Langslow in particular have argued for a higher level of literacy based on the epigraphic evidence, especially inscriptions and potsherds.[23] Collecting all extant and attested inscriptions down to c. 260, Langslow concludes that the early establishment of an epigraphic habit demonstrates a higher level of literacy.[24] Cornell argues that inscriptions on potsherds from sixth- and fifth-century Latium and Etruria point to literacy beyond their highest classes.[25] This supports the level of literacy proposed here. However, as mentioned, public inscriptions had as much significance as symbols as they had as pieces of writing, if not more.[26] Williamson has demonstrated that inscriptions were rarely placed accessibly, and even when they were, the legalese made understanding them difficult.[27] Early public inscriptions demonstrate full literacy early in the Republic, but among only their composers, not the general population. They are evidence for something more like scribal literacy than a general ability amongst the population.

There are also problems with Cornell's argument. Of graffiti scratched near Rome, Cornell asks, 'What kind of aristocrat would scratch his name on a tile?'.[28] It is something of an assumption to believe this could not have been the work of the higher classes (perhaps a younger member?), given the human propensity across all classes to write names on objects throughout history – although this does not prevent Cornell's attribution to those lower in society from being correct.[29] There is also a more serious point here. Name writing is evidence for only semiliteracy, not full literacy. At best, the graffiti provide evidence for a widespread but limited ability to read and write. This is in line with the level of literacy proposed here among the ranks. If the graffiti can be divorced from the highest classes, it supports the notion of more general semiliteracy in Rome from early in the Republic.

In his discussion of literacy in the centurionate under the Principate, Adams argues that complete illiteracy was 'virtually inconceivable' among centurions. He dates this back to Polybius and the second century.[30] Vegetius also stresses the importance of recruiting literate men.[31] The second book of his *De re militari* is notoriously difficult to unpick, with Vegetius mixing elements from the Roman army over several hundred years. Despite this, as the watchword procedure was an old one, it is likely that literacy was an early concern. The process of the Mid-Republican

---

23 Cornell (1991) 25–31; Langslow (2013) 174–8.
24 Langslow (2013) 176–8.
25 Cornell (1991) 22–3.
26 See V:i.
27 Williamson (1987) 162.
28 Cornell (1991) 22.
29 Cf. Baird & Taylor (2011) 5, 7, 11; Huntley (2011) 88.
30 Adams (1999) 126.
31 Vegetius, *Mil.* 2.19.

dilectus would have made deliberate recruitment for literary skills difficult, but identifying the literate and semiliterate may well have had an effect on rank allocation when the recruited reassembled. However, as already shown, this literacy need not be at a high level. Semiliteracy was enough to fulfil the duties of centurion as set out by Polybius and Livy. As there is no evidence for extensive formal military training in the Middle Republic, it is probable that the majority of men both had and were expected to have these skills before enlistment.

Therefore, many soldiers possessed a basic level of literacy capable of dealing with day-to-day written tasks. The watch inspection process demonstrates that a higher level was expected from society's richer members. As it was by and large men from these strata who held magistracies, more complicated administrative tasks were unlikely to fall to those unable to complete them. In a sense (although this strays close to a circular argument), Rome was able to have a military bureaucracy because the skill to create documents existed, thus in the Middle Republic the literary skill existed to create the required military documents. It is impossible to say with the current evidence whether literacy led to the management of the legions in this way or vice versa, or indeed whether they developed in parallel. Thus Harris' picture of widespread illiteracy is not an adequate description of Rome's general state of literacy. Rather, a state of semiliteracy of varying degrees was the norm for the majority of the male population.

ii *Scribae*

In previous chapters, the examination of military administration has been split into two: on campaign and in Rome. However, in this chapter the nature of Roman magistracies means that it is more illuminating to discuss the office rather than its operational location. This section aims to demonstrate that *scribae* (best translated as 'clerks') on the staff of magistrates were responsible for creating many of the documents discussed in the preceding chapters. (The census forms a case apart, and will be discussed in 6:iii.) It is often noted that Republican Rome suffers from the lack of evidence for a 'civil service', although the same scholars admit that it is difficult to believe that many of Rome's systems could have functioned without one.[32] The administration proposed by this monograph sharpens the need for this service, and the work cannot be considered complete without addressing this issue.

Although much work has been done on the roles and appointment of various *apparitores*, including *scribae*, the majority focuses on the

---

32 Toynbee (1965a) 302–3; Purcell (1983) 132; Harris (1989) 155; in particular the census and recording *senatus consulta*. Corcoran (2014) 187 suggests that magistrates relied heavily on their private households for the skills required.

Principate, with its wealth of epigraphic evidence.[33] By the end of the Republic, *scribae* formed an *ordo* of their own, next in status to the equestrians. They formed a permanent pool of state-employed citizen labour associated in particular with the three *decuriae scribarum* of the *aerarium*. There is little doubt that the profession had reached this height by the mid-first century; Cicero was wary of offending the *ordo scribarum*, and the earliest inscription mentioning the *decuriae ab aerario* dates from the same period.[34] However, the evidence for the role and origins of *scribae* in the Middle Republic is much scantier. In Badian's comprehensive list of known named *scribae* from the Republic, only five of 28 predate the first century.[35] Despite this, it is possible to uncover something of the role of *scribae* in the Mid-Republican military sphere.

The earliest known *scriba* is Cn. Flavius. Flavius is noted in the histories because he became curule aedile in 304, causing upset among some of the establishment.[36] Importantly, it his election to office which is noteworthy; his role as *scriba* is presented by Livy, Piso and Pliny as ordinary. The same is true of the other examples of Mid-Republican *scribae*. Claudius Glicia was appointed dictator in 249 but forced to abdicate.[37] The *Periochae* describes him as the worst type of man, indicating that Livy had little positive to say about Glicia. Of the others on Badian's list, L. Cantilius (a pontifical clerk) was executed for fornicating with a Vestal,[38] and Cn. Terentius/L. Petilius discovered the books of Numa on his land.[39] C. Cicereius eventually became praetor and celebrated a triumph on the Alban Mount; it is striking that his former role is mentioned in both the *Fasti Triumphales* and *Fasti Urbisalviensis*, perhaps implying that it weighed against him in the senate's decision.[40] Two unnamed *scribae* were implicated in the trial of L. Scipio.[41] It is the actions or character of the *scribae* which made them noteworthy to historians rather than their profession, but in each example the position of *scriba* is taken for granted as part of the established order. The possibility of anachronism, particularly in the case of Flavius, cannot be entirely ruled out. Nonetheless, the details of Flavius' election and magistracy suggest that on balance *scribae* were an

---

33 On the principate: Mommsen (1893) 383–410; Jones (1949) 38–41; Purcell (1983) 128–9. Purcell (2001) is the main and thorough work on the Republic.

34 Cicero, *Verr.* 3.184; *CIL* 6.1816.

35 Badian (1989) 583–5.

36 Livy 9.46; [Livy], *Periochae* 9; Gellius, *NA* 7.9 = Piso *FRHist* 9 F29; Pliny, *HN* 33.6.17–8.

37 [Livy], *Periochae* 19.

38 Livy 22.57.3.

39 Livy 40.29.2–10; Pliny, *HN* 13.84. The sources disagree on the name of this *scriba*, but the majority of modern scholars follow Herrman (1946) in considering Cn. Terentius the correct attribution, cf. Badian (1989) 586; Purcell (2001) 639; Briscoe (2008) 482.

40 Livy 42.1, 7, 21; Valerius Maximus 3.5.1, 4.5.3; Degrassi (1947) 80–1 ([*qui s*]*criba* [*fuera*]*t*), 338–9 (*qui scr*[*iba*] *fuer*[*at*]); Lange (2016) ch. 3.

41 Livy 38.55.5–7.

established administrative feature by the late fourth century. Although they only occasionally surface in the surviving material, these occasions indicate that they remained an ordinary part of Roman organisation.

That these men were able to fulfil the role of *scriba* suggests that they were more than semiliterate. This implies a good education and thus a richer (if not necessarily high-status) background. Harris suggests that Flavius' rise was based on his ability to take advantage of writing's increasingly important role in Roman administration.[42] This was not the only factor in his rise,[43] but his emergence into the record as the first known *scriba* does point to an increase in the importance of clerks as administration grew more complicated.

Overall, that *scribae* (or ex-*scribae*) could be magisterial candidates, their relatively wealthy backgrounds and their first-century status all suggest that *scribae* fulfilled an important role in the state which required skill and intelligence. Only free or freed men are known to have held these positions, further emphasising their importance. Moreover, although Mid-Republican *scribae* are not well attested, writers felt it worth noting their profession when they were mentioned. Cn. Terentius could have been described as a client without altering the narrative, and others could have been described as *apparitores*. The emphasis on their role as *scribae* suggests that the position was a significant one. Sulla's *Lex Cornelia de XX Quaestoribus* dealt with *scribae* before other *apparitores*, again suggesting that they had higher status and different treatment.[44] Thus, *scribae* held a relatively high position in society, reflecting their wealth and skills.

The discussion can now return to a military theme. Badian has highlighted that it is impossible to recreate the organisation of *scribae* before the first century due to a lack of evidence.[45] Therefore, this discussion will do no more than suggest that the organised systems of *III decuriae ab aerario* and the *ordo scribarum* known to Cicero and his contemporaries developed during the second century.[46] However, it is worth examining what might be said concerning the appointment of *scribae* to magistrates and their work in the *aerarium*.

Badian states that only quaestors and aediles are attested as having *scribae* under the Republic, although he also asserts that praetors, consuls and censors must have used them.[47] However, while the majority of the

---

42 Harris (1989) 155.
43 An association with Appius Claudius Caecus also played a part, cf. Pliny, *HN* 33.6.17; Massa-Pairault (2001) 108–9.
44 *Lex Cornelia de XX Quaestoribus* (trans. Crawford *et al.* (1996) 293f.) 1.6–11.
45 Badian (1989) 598.
46 This is the most that can be said. There is no evidence of these organisations in the third century. A *decuria* is mentioned in connection with Cn. Terentius in 186, but this may be anachronistic (Livy 40.29.10), cf. Purcell (2001) 639.
47 Badian (1989) 598; cf. Purcell (2001) 647–50.

evidence, particularly the epigraphic, focuses on quaestors and aediles, there is evidence that *scribae* were used by more senior magistrates. This leads to a contentious area. There is a school of thought which believes that the more senior magistrates did not have their own *scribae*; instead, the consul or praetor leading an army used the *scribae* of his attendant quaestor. It is due to this association with the highest magistracies, the argument runs, that *scribae* held such a high social status.[48]

Such a view is problematic on several counts. Firstly, the creation of legion lists at the *dilectus* seems impossible on this model. If the consul had recourse only to his quaestor's *scribae*, their presence at the levy could not be guaranteed. The point at which quaestorial *provinciae* were allotted is not entirely clear. Polybius makes no mention of the quaestors being with the legion until he reaches his discussion of pay. It is plausible that the consul could borrow the quaestorial *scribae* before the legion arrayed in the field, but there is no evidence. Alternatively, it is possible that *scribae* from the *aerarium* could be used, especially once a legion-list copy was kept in Rome. At some point a duplicate was produced. However, it was demonstrated earlier that a full list including lines of battle could only have been produced at the first reassembly of the enlisted (I:v). For these events to function smoothly, it would have been eminently sensible (although this of course is no guarantee of reality) for the consul to have his own *scribae*.

More generally, given the amount of documentation proposed by this monograph (legion lists, legion expenses and pay and performance registers), it is unlikely that only two *scribae* were expected to produce and manage the administration of an entire army. With two legions and attendant allies, plus others in the baggage train, this was somewhere in the region of 20,000 men. Further, both the consul and the quaestor using the same clerks might render redundant the precaution of duplicate bookkeeping to guard against corruption. Such a task may well have been devolved to the *scribae*. As the quaestor represented the treasury, the *scribae* on his staff did likewise. The introduction of duplicate bookkeeping would have been of no benefit if both sets of entries were completed by the same men.[49]

On the other hand, it is possible that consuls had access to secretarial assistance in a different form: military tribunes. Each legion had a complement of six who were responsible for the day-to-day running of the army. Two tribunes led the legion on a monthly rotation.[50] What, then, were the duties of the other four? As Suolahti states, service on the general's personal staff is an obvious solution.[51] There is little doubt that the tribunes

48 Badian (1989) 598; contra Jones (1949) 40.
49 See IV:v on legionary bookkeeping.
50 Polybius 6.34.3.
51 Suolahti (1955) 46.

were fully literate. As seen previously, military tribunes wrote detailed instructions for checking the watch.[52] Perhaps a more literate individual was chosen over a less well-educated colleague, but on balance it is unlikely that running the legion – a job with a clear need for literacy – would be entrusted to anyone unable to fulfil this requirement. Moreover, Suolahti highlights that all the known Republican military tribunes had a background of at least equestrian status.[53] He argues that the status of the office was originally second only to the consuls; while this declined when other offices were added as Rome grew, the tribunes' importance as army officers remained the same. This is reflected by the presence of senators and consulars as military tribunes at Cannae.[54] As the legions assumed that cavalry were literate, it follows that men from the same equestrian background were likewise. The relatively high status of *scribae*, from both their financial background and the importance of their position, suggests that such a role with a clerical component was not objectionable to the aspiring elite. It was convenient for a consul to use these well-educated and militarily experienced men as a secretarial staff while they were not commanding the legion.

Military tribunes acting in this capacity provide a possible solution to the difficulties of performing the *dilectus* mentioned above. Polybius' description of the levy suggests that the consuls who would lead the legions were not always present. The selection process centred on the tribunes, allowing for the general's absence.[55] If the military tribunes acted in an official clerical capacity, the legion lists could have been created by them, and additions of battle line made at the reassembly. This removes the need for a *scriba* from the general's staff to be present even if the general himself was not. Further, a clerical role for military tribunes allows them to be involved in keeping the consul's account books. As well as safeguarding against corruption, such involvement would grant those desirous of a political career greater understanding of running a legion. There is no direct evidence, but it is difficult to imagine that a commander would not take advantage of this pool of skilled labour.

There is more to be said concerning *scribae*, however. The military tribunes' assistance in written matters does not demonstrate that consuls and praetors did not have their own *scribae*. Indeed, it might even be considered odd that the consuls did not have a full complement of *apparitores*.[56] The presence and number of lictors, for example, was a prominent mark of status for magistrates with *imperium*.[57] The lack of *scribae*, the highest-status *apparitores*, in a consul's retinue seems strange in a culture where the

---

52 Polybius 6.34.7–36.9.
53 Suolahtu (1955) 55.
54 Suolahti (1955) 37–44; Livy 22.49.16.
55 Polybius 6.20.
56 Cf. Purcell (2001) 648f.
57 Livy 1.8.3; Dion. Hal., *Ant. Rom.* 5.74.2.

physical presence of a retinue and clients was an important indicator of an individual's status and influence. That a consul might have a smaller retinue than his quaestor, a much more junior magistrate, seems unlikely. Moreover, even if the military tribunes were involved in written duties, men dedicated to this work without distraction can only have been beneficial. The military tribunes available each month would change, breaking continuity, and all had other responsibilities in the camp.[58] The best conclusion is perhaps that clerical tasks such as creating the legion lists and writing watchwords and any other written orders were within the normal and expected arena of the military tribune, but other written work, such as account books, was given to *scribae*. Military tribunes had an important clerical role within the legion, but as part of their command role.

Further, there is evidence of *scribae* attached to magistrates other than quaestors and aediles. Pliny notes that Cn. Flavius had been the *scriba* of Ap. Claudius Caecus.[59] Aulus Gellius and Livy, both following Piso, say that he was at the election as a *scriba* in the service of an aedile.[60] As Jones points out, the two attributions are not inconsistent.[61] If *scribae* were already organised into pools of *decuriae*, Flavius could have been selected to work for them in different years. Even if – as on balance seems more plausible – this labour pool had not yet been formalised, it is not surprising to find a skilled *scriba* retained on more than one occasion. The position held by Caecus when Flavius was his *scriba* is not mentioned. Nonetheless, as Caecus was censor in 312, it is likely that he was of a higher position than aedile when Flavius was his *scriba*.

Flavius is not the only *scriba* who may have been in the service of a high-ranking magistrate. C. Cicereius, who himself became praetor, may have been a consul's *scriba*. Valerius Maximus states that Cicereius was the *scriba* of P. Cornelius Scipio Africanus.[62] Scipio's position at the time is unclear, but given his swift political rise, consul or proconsul is not an unreasonable assumption.[63] Moreover, Broughton suggests that Valerius is mistaken: the Scipio in question was in fact L. Scipio, Africanus' brother.[64] If Broughton is correct, there is perhaps more which can be learned about Cicereius and Republican *scribae*. During the trials of the Scipiones, two *scribae* were among those on the same charges as L. Scipio, although they were acquitted before Lucius' trial.[65] Once again it is unclear whether these

---

58 Polybius 6.19–42.
59 Pliny, *HN* 33.6.17.
60 Livy 9.46.2; Gellius, *NA* 7.9.2 with Oakley (2005) 603.
61 Jones (1949) 38.
62 Valerius Maximus 3.5.1, 4.5.3.
63 The only other magistracy Scipio held was the curule aedileship in 213, Livy 25.2.6–7.
64 Broughton (1951) 406 n.2.
65 Livy 38.55.5–7. Livy implies that the two *scribae* and an *accensus* were added to the charge sheet in order to emphasise the sense of a conspiracy, presumably because their role would have involved keeping the books which noted the 'bribes'.

men were attached to Scipio or to his quaestor. However, that Livy specifically notes two *scribae* rather than just *scribae* implies that there were more with Lucius' army. As Cicero states that quaestors were attended by two *scribae*,[66] these other unindicted *scribae* were attached to either the quaestor or another magistrate. The general is the obvious candidate. Thus the evidence suggests that consuls did have their own *scribae* on their staff. Cicereius may have been one of these men. It is possible that he was retained as a *scriba* in a personal rather than official capacity after Scipio had held office, but this does not affect the conclusions here; as will be seen, the division of public and private is unlikely to have been as strict as the *decuriae* later made it.

The cases of Flavius and Cicereius may also provide some insight into the organisation of *scribae* prior to the establishment of the *ordo scribarum* and the *III decuriae ab aerario*. In describing both Flavius and Cicereius as *scribae* in association with a particular individual, Pliny and Valerius indicate a current and thus lasting tie to that individual.[67] Flavius' political activities suggest he had a longer association with Caecus than just a year. In the case of Cn. Terentius, the most said is that Q. Petilius, then a praetor, was a quaestor when he first recognised the former's abilities.[68] Cn. Terentius may represent the first evidence of the development of *decuriae*, but with the others he nonetheless demonstrates a principle of scribal organisation. *Scribae* maintained a relationship with a senator beyond a single year. The relationship during the fourth and third centuries may have been more akin to that of the earliest quaestors with their consuls: personal selection by the magistrate creating a relationship more personally charged than the later system of allocation. It is easy to see that greater objectivity from *scribae* would have become increasingly desirable as the Republic and its influence increased, much as it had been with making the quaestor, the state's financial representative, more removed from the general on whom he was to keep a check.

It is possible that the *scribae* of quaestors and aediles are more prominent in the evidence due to the association of these two magistracies with the *aerarium*. By the first century, the *aerarium* had three *decuriae* of clerks associated with it. These *scribae* had a permanent role in the treasury undertaking the treasury's business, overseen by the quaestors of that year.[69] There was a limit to the authority of these quaestors over the *decuriae*. Cato the Younger was able to prevent a corrupt *scriba* from working under him,

---

66 Cicero, *Verr.* 2.3.182.
67 Pliny, *HN* 33.6.17 – *ipse scriba Appi Caeci*; Valerius Maximus 4.5.3 – *scribam C. Cicereium*.
68 Livy 40.29.2–10.
69 Plutarch, *Cato Min.* 16.

but not to remove him from the *decuriae* entirely.[70] It is unclear exactly when this system came into being, but something must be said regarding the organisation of the *aerarium* prior to the establishment of the *decuriae*.

The development of the permanent *decuriae ab aerario* indicates a need for a permanent clerical staff in the *aerarium*. This in turn implies a substantial amount of paperwork, especially as the *aerarium* dealt not just with military documents but with financial papers and *senatus consulta*. Such a quantity of work, in composing, copying and archiving, required a skilled staff before it reached the volumes of the Late Republic. Who, then, were these individuals? The most obvious answer is that they were the *scribae* on the staff of the supervising quaestors. If scholars are correct in supposing that the number of quaestors was increased to eight in 267,[71] this may be the point at which two urban quaestors were appointed. Such a development implies an increasing volume of work to be undertaken by the *aerarium* even before the major military expansions of the late third and second centuries. Even if the entire *aerarium* staff consisted of the quaestor and his two *scribae*, doubling this was a significant increase in the manpower available. Further, in the earlier period of the Middle Republic when campaigns were largely limited to the summer months, the *scribae* of other quaestors may have been available for use in the *aerarium* during the winter.[72] Indeed, as the return of armies and their documents probably occasioned more work in the *aerarium* through the updating of various military and financial records, having the men on hand who had created the documents on campaign would have been an advantage.

It is possible that urban quaestors may have chosen their *scribae* for the year from those experienced at working in the *aerarium*, creating a de facto permanent staff which eventually became the *decuriae ab aerario*. Cato the Younger's struggles demonstrate that the quaestor's duties could be overwhelming for a young magistrate in the Late Republic without considerable preparation.[73] On the other hand, in the Middle Republic a new quaestor may well have served as a military tribune during his required military service. This would have given him some experience of accounting, even if not to the level required for the *aerarium*. Nonetheless, the presence of at least one experienced *scriba* on his staff would be a great help to a quaestor taking on the *aerarium*.

The quaestors may not have been the only magistrates who could bring *scribae* to the *aerarium*. Livy records that in 202 the *scribae* of curule aedile L. Licinius Lucullus were caught stealing from the *aerarium*.[74] It is possible

70 Purcell (2001) 654.
71 E.g., Harris (1976); Erdkamp (2007) 107.
72 Providing that the *scribae* themselves did not need to return home to farm.
73 Plutarch, *Cato Min.* 16.
74 Livy 30.39.7.

that this treasury is the otherwise unknown aediles' treasury mentioned by Polybius.[75] However, Livy gives no indication that this *aerarium* was different to the *aerarium Saturni*. On the assumption that this is the *aerarium Saturni*, the theft suggests that the curule aediles' *scribae* assisted with treasury work. Livy presents the theft as timed with the games given by the aediles, indicating that the *scribae* perhaps hoped that the loss of the extra funds would go unnoticed at a time of great expenditure.[76] More significantly, it suggests the *scribae* had easy access to the *aerarium*. It appears, therefore, that quaestorial *scribae* were not necessarily the only *scribae* employed in the *aerarium*.

The growing importance of *scribae* during the Middle Republic is reflected in the granting of a Temple of Minerva to them as a quasi-headquarters in 207. Festus mentions this dedication alongside an explanation of the term *scriba*, which he asserts had not yet separated into *librarius* (copyist) and *poeta* (poet) but covered both.[77] The gift of a temple indicates that clerical roles were increasingly valued, answering a need closely related to the army's organisation and recording manpower. However, this aspect is often overlooked in favour of the *poetae* due to the temple dedication's connection with Livius Andronicus. Andronicus' dates are disputed, but it is unlikely he lived to 207.[78] Rather, he was probably tutor to M. Livius Salinator (cos. 207), having been enslaved by Livius senior in 272 at the fall of Tarentum. If Andronicus was a *scriba*, rather than specifically a *poeta*, it is possible that he worked in the role's more practical side following his emancipation sometime between 272 and 250. M. Livius senior served as decemvir in 236 and may have been an ambassador to Carthage in 218.[79] While these offices did not officially have *apparitores*, it has been shown that *scribae* had a closer personal relationship with their magistrates prior to the *decuriae*. Therefore, Andronicus could have performed a secretarial role for M. Livius, even if he was not a *scriba*, as a formal *apparitor*. The lack of distinction between clerk and poet in the language indicates that this distinction was yet to develop. Thus the Temple of Minerva was for all those who employed written skills professionally, encompassing the *scribae* dealing with military administration.[80]

Thus *scribae* serving on magisterial staff, both in Rome and in the field, carried the bulk of the responsibility for producing, maintaining and preserving the administration which documented Rome's military. Within the legion itself, the military tribune carried some of this burden, particularly during legion formation. It was on the legion lists that the rest of military

75 Polybius 3.21–6.
76 Livy 30.39.6.
77 Festus 446–7L with Purcell (2001) 644.
78 Mattingly (1957) 161–2; contra in part Beare (1940) 12–5.
79 Livy 21.18.1 with Beare (1940) 14; Broughton (1951) 223.
80 This gathering and collective consideration of *scribae* may have been the first step towards the *ordo scribarum*.

administration undertaken by *scribae* in the field, especially pay and per-formance registers, was based. As the Middle Republic progressed and Rome expanded its influence and military activity, this clerical role became increasingly crucial, as the increased number of quaestors illustrates. By the middle of the Hannibalic War, the importance of a dedicated, professional, literate class (although perhaps alien to the Roman aristocratic ethos of unpaid service) was recognised and began to develop into the more re-cognisable form found under the Principate. The *scribae* both assisted the development of Rome's military ambition and were made necessary by it.

iii Recording the census

The ancient evidence for recording the census during the Middle Republic is both scant and apparently contradictory. Despite this, it is necessary to investigate who created and maintained the census, the central record of military manpower. This section examines the state of the evidence in order to suggest possible solutions.

A central point is whether the census roll was updated only every five years during the census or whether a permanent staff was housed in the *atrium Libertatis* to make yearly alterations. The latter is attractive to several scholars, who argue that deaths and the movement of men through different age groups needed to be noted to keep the record of manpower up-to-date.[81] This conclusion has its merits: the Roman concern for manpower and accurate recording, as well as political need, suggests such a system was desirable; and, more significantly here, failure to update the age classes could have denied Rome access to its youngest and fittest citizens, who turned 17 between *lustra*. However, no modern scholar has produced evi-dence to support this hypothesis. This discussion hopes to demonstrate that such a conclusion does not find support in the surviving evidence.

It is worth beginning with the only *scribae* yet to be discussed, those of the censors. The censorship is the other high magistracy with evidence of accompanying *scribae*. Livy states that at the establishment of dedicated censors in 443, it was decreed that they have *scribae* to assist them with their work.[82] These censorial *scribae* are otherwise absent from the record of the Republic, apart from Varro's passing notice of their involvement in closing the *lustrum*.[83] Particularly striking is Livy's failure to mention them during the cessation of the 169 census.[84] Nevertheless, something can be gleaned concerning the role of the *scribae* from these brief mentions. Firstly, the two passages in combination demonstrate the importance of *scribae* to

---

81 Bourne (1952) 133; Toynbee (1965a) 302, 449; Hin (2008) 214–8.
82 Livy 4.8.4.
83 Varro, *Ling.* 6.87.
84 Livy 43.16.13.

the census process. They were involved from the outset and received purification alongside the censors and other magistrates who took part in the
lustrum ceremony.[85] Secondly, Livy indicates that the written aspect of the
census, turning spoken declarations into physical records, was always a key
part of the censors' work. Work on this scale required the support of skilled
clerks, a need recognised from the very beginning. Coupled with their involvement in closing the *lustrum*, the evidence indicates that *scribae* had a
crucial role in the census, a role recognised and honoured by the state. Such
a prominent position leaves little doubt that they were involved in the
census process throughout the Republic despite a lack of other evidence.

It must then be questioned what role the *scribae* performed, as they are
not found elsewhere in census descriptions. The answer may be a simple
one. It can be inferred from Livy that the *scribae* were initially engaged to
assist the censors in recording the oral declarations of Rome's *patresfamilias*. It is well established that census declarations were given to
officials known as *iuratores*.[86] These were free men, like all known *scribae*,
who assisted the censors in just the manner described. In the Ahenobarbus
relief, the man to the far left listening to the toga-clad citizen's declaration is
also recording it on a large *tabula*.[87] He wears a toga, indicating that he too
is a citizen. It is possible that he depicts the censor himself, but he is
commonly identified as a *iurator*.[88] The *iurator* needed to be fully literate to
record the new declaration as well as read that of the previous census and
make any necessary changes. Whether that happened at the point of declaration or at a later stage of compilation is not important here. Thus the
roles of the *scriba* and *iurator* appear to overlap entirely; indeed, the desirable qualities were the same in both.

However, the possibility that these were discrete positions must be addressed. It is not implausible that the *paterfamilias* made his declaration to
one man but it was recorded by another. Livy's description of the censorship's beginning suggests that the *scribae* were to record declarations, but it
is less clear whether they were addressed when the declaration was given.
The censors themselves fit this role better. On the other hand, if the *scriba*
and the *iurator* were not the same man, who is the *paterfamilias* in the
Ahenobarbus relief addressing? On this hypothesis, the seated figure must
be the *scriba*. Given the *iurator*'s presence over *scribae* in written sources, it
would be surprising if the relief's designer not only gave the recorder a more
prominent role than the *iurator* by placing him in the foreground, but
omitted the *iurator* entirely. Moreover, the *paterfamilias* appears to be

85 Varro, *Ling.* 6.87 – *censor < es > scribae magistratus murra ungentisque ungentur.*
86 Livy 39.44.2; Cicero, *Leg.* 3.7; e.g., Mommsen (1894) 37–8; Suolahti (1963) 34;
   Northwood (2008) 258.
87 See IV:i n.37.
88 E.g., Nicolet (1980) 87; Torelli (1982) 10.

addressing the seated figure. Coupled with the failure of extant sources to mention *scribae* and *iuratores* together, the relief suggests that the two were the same.

If the *iuratores* and *scribae* were one and the same, it is likely that each censor had more than two *scribae*. This need not be a problem, however. Cicero only states that quaestors had two *scribae* each.[89] Other magistracies are not mentioned. In the case of censors in particular, it is likely they could employ as many men in this role as necessary for the census' timely completion. For example, they were able to send legates to the legions in 204.[90] As already discussed (II:ii), this was unprecedented but evidently within their power. There is no evidence of a set number of *iuratores*; this does not mean that there was no customary figure, but it does suggest a flexibility allowing the censors to fulfil their duties unhindered by a lack of manpower.

The *scribae/iuratores* provide an insight into the keeping of census records. Both *scribae* generally and *iuratores* in the census were only present in the retinue of a magistrate for the period of office. They ceased to have responsibility for the census documents following their ritual deposition. Mid-Republican *scribae* had a connection to the individual rather than the magistracy; *iuratores* had no declarations to record once the *lustrum* closed. Thus these clerks were no longer in the service of either censor or census. This suggests that there was no one available to make any changes to the census between *lustra*, providing an obstacle to the opinion that there must have been a permanent clerical staff keeping the census updated.

Livy's description of the census disruption in 169 may, however, overcome this obstacle. Livy states that during the disruption the censors closed the *atrium Libertatis*, sending away the public slaves who were present there.[91] He makes no mention of *scribae* or *iuratores*, but closing the *atrium* nonetheless suspends the census' operation. This suggests that it was the public slaves who were responsible for collating the declarations taken by *iuratores* into the completed census list. Further, slaves could not be dismissed until their service was required again. This does not rule out the slaves having alternative employment between *lustra*, nor does it prevent a new group of slaves being used in the future; literacy was the only essential skill. Nonetheless, it suggests that there was an available pool of labour familiar with the operation of the census and the archives themselves. They could be used to make the changes necessary to keep the census documents up-to-date. If the public slaves were used in this way, then continuous work on the census could have occurred.

89 Cicero, *Verr.* 2.3.182.
90 Livy 29.37.5–6.
91 Livy 43.16.13.

However, there are several problems with this conclusion. Chief of these is Livy's failure to mention the *scribae/iuratores*:

> *censores extemplo in atrium Libertatis escenderunt et ibi obsignatis tabellis publicis clausoque tabulario et dimissis servis publicis negarunt se prius quidquam publici negotii gesturos, quam iudicium populi de se factum esset.*
>
> (Livy 43.16.13)

At once the censors went up to the *atrium Libertatis* and there having sealed the public tablets and closed the record room and dismissed the public slaves they refused to do any work of public business until the judgment of the people concerning themselves had been made.

The passage is specifically concerned with the *atrium Libertatis* itself. The censors were able to suspend their public business by closing the archive room and the tablets holding census information. The *iuratores* were presumably absent because their work was conducted at the *Villa Publica* on the *campus Martius*. It is probably from there that the censors 'went up' to the *atrium*. The most probable explanation for the lack of *iuratores* is that they were dismissed before the censors proceeded to close the *atrium*. Even if the declarations themselves were temporarily stored in the *Villa Publica* (or the *aedes Nympharum*, if it yet existed), closing the *atrium* with its archive prevented the compilation of the full census document, thus suspending
the censors' work. Livy's account does not demonstrate that public slaves were responsible for writing documents, only that they were present in the building.

What, then, was the role of the public slaves? Suolahti and Briscoe have followed without comment Mommsen's assertion that the slaves were responsible for all the census work.[92] However, the only evidence they provide is this very passage. The slaves may have formed a permanent staff in the *atrium Libertatis*, but one of caretakers rather than clerks is more likely. Mouritsen points out that slaves and freedmen were not used by Rome in any recorded military capacity.[93] The 8000 slaves who served in the Hannibalic War (eventually earning their freedom) stand out as an exception to this.[94] It is possible that this prohibition extended to creating Rome's manpower register (see II:i). More significantly, slaves were manumitted through inclusion in the census.[95] Buckland argues that this was a

---

92 Mommsen (1893) 337; Suolahti (1963) 34; Briscoe (2012) 445.
93 Mouritsen (2011) 73.
94 Livy 22.57.11; 24.34.3–9.
95 Cf. Buckland (1908) 440.

recognition of free status by registration in a list of citizens, not itself an act of manumission: 'the censor is recording the fact that the man is a *civis*, not making him one'.[96] It is highly unlikely that a slave would be entrusted with the compilation of a list which could in effect grant him his freedom.[97] It is always the censor who is mentioned as noting manumission in the ancient sources, but the foregoing considerations suggest that allowing slaves a clerical role in the census was potentially worrying.

The importance of having citizen *scribae/iuratores* emphasises the distinction between the permitted roles of slave and free. The census was of central importance to the state's organisation, both politically and militarily. The use of only free and freed men in clerical, administrative roles functioned as a precaution against involving in sensitive work those without a stake in it. Livy mentions the public slaves in the *atrium Libertatis* because they performed a caretaking role, perhaps moving documents from the *campus Martius* to the *atrium* as part of this role. For the censors, dismissing the public slaves was a symbolic as well as practical act of ceasing operations, because, like sealing the tablets and closing the archive, it marked a halt in normal business.

Such a division between free clerks and slave caretakers is evident later in the Republic. It is not unproblematic to use evidence of a later period, but it is plausible that this period's arrangements originated earlier. Houston highlights that Cicero did not have a slave dedicated to working with his library. Instead, the slaves who cared for his library also had other tasks about the house. His friend Atticus did have such dedicated slaves, but these were *librarii* (copyists) rather than *scribae* (clerks).[98] Cicero refers to Atticus' library slaves as *librarioli*, 'little copyists'. These slaves had the skills of archivists; Cicero borrowed them for tasks including mending damaged works. However, there is no indication that they performed any of the tasks associated with Mid-Republican *scribae*. Cicero engaged a free scholar to organise his library.[99] This labour division suggests that, here as with the *atrium Libertatis*, it was free men who were considered skilled and responsible for organising and creating documents, while slaves were considered suitable caretakers.[100] The tasks undertaken by the *librarioli* do suggest, however, that caretakers were trusted to work with documents for the purposes of preservation. This tallies with the apparent role of the *atrium Libertatis*' public slaves.

---

96 Buckland (1908) 441, followed by Mouritsen (2011) 11.
97 Cf. Mouritsen (2011) 20–1 on negative attitudes towards slaves.
98 Cicero, *Att.* 4.4a.1; Houston (2002) 141–2, 147.
99 Cicero, *Att.* 4.8.2, *Q. Fr.* 3.4.5.
100 Tiro stands out as a counterexample, but he had a close and lasting relationship with Cicero. Neither the slaves who mended Cicero's books nor the public slaves of the *atrium Libertatis* had such a relationship with their presiding citizens.

Perhaps more significantly, Houston suggests that after its rebuilding in 28 the *atrium Libertatis* was staffed by public slaves.[101] Unlike many libraries formed under Augustus, the *atrium* was not part of Augustus' household.[102] Other libraries were staffed by Augustus' household slaves, as would be expected of his property.[103] However, the *atrium* was a rebuilt public building, as well as the first such library. A staff of public slaves would be expected. If the Augustan *atrium* was staffed in this manner, it may well have been carried over from the Republican operation. As librarians or archivists, these slaves were not responsible for the generation of content, only its preservation, which could include copying.[104] It is probable that Mid-Republican public slaves engaged in the *atrium Libertatis* had a similar role and should be equated with Atticus' *librarii*.

It remains possible, however, that slaves were involved in the census in a more clerical fashion but are not found in the sources because the ancient authors assumed their readers were familiar with their role. The lack of evidence for either position means it is not possible to be conclusive. It is possible, for example, that *iuratores* were a screen, with unmentioned slaves doing the actual clerical work.[105] However, there are several objections. It was already demonstrated that a citizen, most probably holding the position of *scriba/iurator*, recorded the declaration of a *paterfamilias*. The clear depiction of a citizen recorder on the Ahenobarbus relief should also not be overlooked in favour of unattested slave *scribae*. Moreover, the inclusion of *scribae* in closing the *lustrum* alongside senior magistrates suggests that they had a major role.[106] All the extant evidence for the role of the free citizen *scriba/iurator* coupled with the known role of later slave *librarii* in archives is more persuasive than the unattested possibility of slave *scribae*.

Aside from the censorial staff, there is other evidence to suggest that the census was not updated between *lustra*. Firstly, the nature of the *lustrum* itself: as has already been discussed (II:v), closing the *lustrum* was a solemn event in which the physical embodiment of the citizen/noncitizen divide was enshrined. It seems implausible that changes to the census list could be made after this ceremony without the supervising authority of the censors. Moreover, as the public slaves were the only permanent staff at the *atrium Libertatis*, the same objections to their involvement in this process apply, if not more so due to the lack of a magisterial overseer. Quite when changes made by the censors came into effect during the census period has been debated, but they were certainly in force after the *lustrum* ceremony.[107]

---

101 Houston (2002) 157.
102 Pliny, *HN* 7.30.115.
103 *CIL* 6.4435, 6.8679, 14.196.
104 Cf. Purcell (2001) 641 f.
105 Andrew Fear (pers. comm.).
106 Varro, *Ling.* 6.87.
107 E.g., Buckland (1908) 441.

During the census period, changes were enacted through the moral power of a censor elected to his authority by the people; it is improbable that they could be rendered by an unelected inferior, let alone a slave, at other times. The lack of an authority figure between the sacred dedications of *lustra* suggests that the lists were not updated between censorial magisterial periods.

Secondly, the emergency levy, the *tumultus*, suggests the same. The necessity of holding a *tumultus* indicates that aside from a state of emergency, the lists held by the senate were not entirely accurate. As discussed previously (II:iii), if Rome wished to mobilise all those of fighting fitness, a *tumultus*, not lists generated from the census, was the way to achieve this. This overcame any weaknesses in census taking, including any changes which had occurred since the most recent census. Thus, the use of *tumultus* in the Middle Republic contributes to a picture of census lists which were not entirely accurate.

Thirdly, the gap between census periods. Suolahti suggests that holding the census every five years became an established custom because after five years the previous list became unworkable for organising military and political matters.[108] He does not provide any evidence to support this hypothesis, but it is plausible. Dionysius of Halicarnassus explicitly states that the census declaration included the ages of children.[109] The *tabula Heracleensis* also states that the names of all in the family are to be given.[110] As Northwood points out, the inclusion of ages allowed the list to operate over the five years until the next census.[111] Recording ages indicates that even though the census itself was unaltered, updated lists could be generated from it, such as the *tabulae iuniorum*. More importantly, Dionysius' emphasis on recording the ages of children (rather than just adults for distribution into the correct age class) indicates that accurate *tabulae iuniorum* were a consideration of the censors. Dionysius wrote at the end of the first century, a period in which army recruitment had all but ceased to be based on Polybius' *dilectus*. That he records the inclusion of ages in the census suggests that this was carried over from an earlier period. The same can be said of the *tabula Heracleensis* (2:i). Thus it appears that recording children's ages in the census was a deliberate act aimed, at least in part, at allowing into the *iuniores* men who turned 17 between *lustra*. It follows that movement into the *seniores* and *senes* could also be calculated.

The chapter can now move towards a conclusion. The census records were generated during the census period under the censors' authority. The declarations of *patresfamilias* were recorded by *iuratores*, who were in all

---

108 Suolahti (1963) 32.
109 Dion. Hal., *Ant. Rom.* 5.75.5.
110 *Tabula Heracleensis* ll.146–7.
111 Northwood (2008) 258.

likelihood the *scribae* allotted to the censors among their *apparitores*. These declarations were collated into the final census roll, essentially an updated version of the previous census list. The completed list was ritually deposited in the *atrium Libertatis* in a procession where *scribae* held an honoured place. Following the closure of the *lustrum*, the lists stored in the *atrium Libertatis* under the care of public slaves could not be altered. However, they still provided the information required to keep the state functioning relatively efficiently. The census records themselves could not be altered, but it was possible to generate up-to-date *tabulae iuniorum* each year for the *dilectus*. This arrangement ensured that the religious strictures surrounding the census were adhered to but the lists remained secularly workable. Coupled with the legion lists generated by military tribunes at the levy, and records of citizen/soldier pay and performance kept by the quaestor and his *scribae* on campaign based on the legion lists, it was possible to understand the full state of Rome's military power using these documents, excluding only those who had failed to register in the census.

It is only possible to speculate on who generated these derivative lists, particularly the *tabulae iuniorum*, from the census each year. It is possible that the public slaves were permitted to do this work under instruction from the senate; it was after all only selective copying of the original census. Alternatively, the *scribae* who attended the consuls may have been responsible for drawing up the *tabulae iuniorum*. If these *scribae* were chosen around the time of the consular elections, they had several months to draw up the lists before the senate made decisions about recruitment and deployment. On balance, because the *scribae* were free citizens and known to be skilled at drafting documents, it is more plausible that the consuls' *scribae* were responsible for generating the *tabulae iuniorum*.

Thus, the organisation of the Roman state and military provided the means for generating the military records proposed in this monograph. In the Middle Republic this had yet to become the organised clerical system known from the first century and the Principate. Nonetheless, throughout the Middle Republic Rome was able to utilise the abilities of the fully literate in order to manage its armies. This came both in the form of professional or semiprofessional clerks, the *scribae*, and from an assumption of some literacy in all serving citizens. For those from the higher classes – consuls, quaestors, cavalry and military tribunes – this assumption was of full literacy, but even for the rank and file semiliteracy was expected. This ability both fuelled the growth and complexity of the empire as its legions and their scope grew and was in turn made necessary by it. As the growth of the number of literate figures and the increased recognition of them by the end of the third century demonstrates, the Mid-Republican army was increasingly fuelled by bureaucracy.

Conclusion
The Mid-Republican origins of Roman military administration

The foregoing discussion demonstrates that the development of Roman military administration into the complex systems found in surviving Imperial documentation cannot be attributed to Augustus. Rather, military bureaucracy developed in tandem with Rome's martial expansion across the Mediterranean. As the scale and scope of these campaigns increased, the methods and mechanisms of bureaucracy grew and changed to cope with new requirements. Operational and administrative abilities developed through a process of mutual reinforcement. This process is particularly recognisable in the changes in Rome's power, influence and operations which occurred during the Middle Republic. The development from the fourth century to the mid-second can now be summarised.

Roman military administration has its origins in the initial establishment of the census as a record of manpower, traditionally attributed to Servius Tullius.[1] By the mid-fifth century this document had become important enough for the senate to create a magistracy, the censorship, whose sole duty was its upkeep.[2] However, it is only with the Middle Republic that the more detailed nature of this administration becomes apparent. At some point during the fourth century there was a change from the simpler phalanx-like army to the more complex manipular legions familiar to Polybius.[3] The increased complexity of recruiting these legions necessitated more organisation behind the scenes. On cue, in the late fourth century *scribae* (clerks) first appear in the works of ancient authors, presented as already well established.[4] This indicates the emergence of a system of documentation and bureaucrats supporting the state's leadership. As one of the major functions of this leadership was martial, it is unsurprising that military administration also becomes more apparent from this period. It is precisely the development of this administration from its humble origins

1 Livy 1.42.5–44.3.
2 Livy 4.8.2–4.
3 Livy 8.8.3–14 with Gilliver (1999) 15.
4 Livy 9.46; [Livy], *Periochae* 9; Gellius, *NA* 7.9; Pliny, *HN* 33.6.17–8. See VI:ii.

that can be seen throughout the Middle Republic from the late fourth century.

In the third century, as throughout the Republic, the census remained the central document key to military administration. These records were compiled during the 18-month census period from the declarations of each *paterfamilias* given to the censors or their assistants.[5] Following the ritual *lustrum* ceremony at the end of the census period, these records were then stored in the *atrium Libertatis* on the Capitoline (the 'headquarters' of the censors, for lack of a better term) under the care of public slaves.[6] From the census records the *tabulae iuniorum* could be generated each year.[7] Each *tabula* contained a tribal list of all those liable and able to serve in the legions. Anyone exempt from the levy, for reasons such as injury, *emerita stipendia* or agricultural labour requirement, was not included on the *tabulae iuniorum*, as the exemption was granted by the censors during their work.[8] The expedient of including the ages of all those declared on the census allowed new *tabulae* to be generated each year without requiring an entirely new census or emendations to the ritually deposited version.[9] The ability of the consuls to grant an exemption at the point of the levy meant that any shortcomings in the census process, such as a citizen reaching *emerita stipendia* between census periods, could be overcome.[10]

Short campaigns dominated the late fourth and early third centuries, resulting in a relative ease of administration. It was rare for a citizen to be away from Rome for long enough to miss the 18-month census period; registration by *paterfamilias* approximately halved the number overlooked in this event.[11] Pay was entirely decentralised, being collected and distributed by the *tribuni aerarii* with no involvement from the legion's magistrates.[12] Outside the First Punic War, reinforcement is absent from the extant record before the Hannibalic War. In the late fourth and early third centuries, shorter campaigns and smaller armies suggest that it was unlikely that Polybius' maximum service term of six years was met unless voluntarily.[13] In all, the nature of warfare in this period required much less in the way of complex and parallel administration between the legion in the field and Rome. The usual return of consuls and their armies to the city with the legion lists at the end of the years allowed Rome to remain the administrative hub of the military with relative ease. Carrying documentation

---

5 See II:i.
6 See V:iii and VI:iii.
7 See VI:iii.
8 See I:iv.
9 See VI:iii.
10 See I:v.
11 See II:ii.
12 See IV:ii.
13 Polybius 6.19.2.

with the campaigning forces was necessary for successful tactical operation, but it remained fully reliant on the hub to which it ordinarily returned yearly. The regular contact between Rome and its legions did not require the parallel systems later developed to keep both parties as up-to-date as possible.

The system developed by the late fourth century continued to operate into the late third century largely unchanged, as the built-in safety measures absorbed the pressures of Rome's military ambitions. The only effect significant enough to emerge in the historical record, and that only obliquely, was the centralisation of *tributum*.[14] Whilst still collected by the *tribuni aerarii*, the distribution of military pay was now handled through the *aerarium* as increasingly long, distant campaigns rendered the fourth-century system inoperable. For this new system to work, the individual pay records known to Polybius were introduced.[15] These could encompass the complications arising from replacing lost equipment, but short campaigns would have resulted in much less differentiation than might later be found in the service records of men serving continuously for six years or more.[16] This change, perhaps more than any other, laid the groundwork for future bureaucratic developments.

The changes to the scale and scope of Roman warfare wrought by Hannibal's invasion of Italy forced further development of these administrative mechanisms in order to continue the efficient application of Rome's military power. The systems of the fourth and earlier third centuries remained in place, but additions and changes were made. In 204 the censors were forced to act more proactively, as completing an accurate census had become almost impossible.[17] The legates dispatched by the censors to the legions in this year highlight that while the senate may have known how many legions were in the field and their approximate standing strengths, they did not know who made up the legions. It is probable that it was the need for this mission which resulted in copies of the legion lists being kept in Rome as well as with the legions. Censors could then cross-check those missing from the current census with those on service, allowing the census to be more accurately maintained.[18] As the central record of Rome's manpower, the census remained key to military administration, developing to cope with the new demands of extended campaigns further afield.

The changes heralded by Hannibal also made the service term limits more significant as longer service became more regular. The effect of this pressure

---

14 It is possible that there were further changes recorded by Livy in his lost second decade. However, the processes apparent during the Hannibalic War do not appear different in any substantial way to those of a century previously.

15 See IV.

16 See I:iii.

17 Livy 29.37.5f.

18 VI:iii.

was ultimately realised in 169: the censors introduced a new oath which implied that enlistment was tantamount to embarking on a campaign which would last for Polybius' entire normal service term.[19] In a sense, this simplified the system. It perhaps suggests that it was no longer necessary to record years served, only whether a man had served or not, in order to prevent extra, potentially aggravating, calls on citizen-soldiers. On the other hand, as service terms remained (at least in theory) the qualification for holding a magistracy, it is probable that some mark of years served remained in the census declaration.[20] More importantly, the development of 169 (which appears to regularise an existing reality in second-century Spain at least) demonstrates that Rome understood that overly long service terms, especially continuous service, could potentially cause mutinous behaviour. Such behaviour was detrimental to military effectiveness and thus to be avoided. The censors' actions demonstrate that Rome both was aware of and worked to mitigate the threat.[21]

Such concerns can also be seen in the final major change to occur as a consequence of the war in the late third century. As longer campaigns became the norm, so did the changing of commanders. It was not unheard of in the third century, but it became almost a yearly occurrence in the second century.[22] When new commanders took their place they conducted a *lustrum*, a military review. This informed the incoming general of the full extent of his new command, rectifying any mistakes in enumeration. This demonstrates recognition of human fallibility: however conscientious the administration of the army, extended periods in the field could lead to errors. The *lustrum* acted as a safety net, catching these mistakes. It indicates that maintaining accurate information on the army was a priority.[23]

The wider use of strategies such as a new commander's *lustrum*, coupled with the introduction of procedures like keeping a legion list in Rome, indicates that the Romans had an active desire to keep their armies as efficient as possible. It may have been conservative in nature, but the senate was also pragmatic, developing the administrative mechanisms in order to account for the pressures caused by longer campaigns further afield. Change was piecemeal, and relatively slow, but nonetheless present. That these developments of the late third and second centuries took place at all demonstrates both the importance of administration to the function of the Roman military and, more significantly, Rome's recognition of this importance. The organisation of the Mid-Republican military was not ad hoc confusion, but a gradually developing system of some complexity.

---

19 Livy 43.14.5–6.
20 Xolybius 6.19.4
21 See II:ii
22 Demonstrated by Livy's yearly lists of appointments and commands, e.g., 39.45.1–8.
23 See III:v.

To summarise, then: by the mid-second century, Roman military administration had reached its zenith in the Republic. The census formed the central document, containing for its military purposes a record of Rome's men, their ages, property qualification, service record and any exemptions. This document was compiled from *paterfamilias* declarations recorded by the censors and their subordinates and cared for after completion by public slaves in the *atrium Libertatis*. Each year, *tabulae iuniorum* were generated from it, probably by *scribae* attached to the generals. Under the senate's instruction, military tribunes (for new legions) or consuls (for reinforcements) used these *tabulae iuniorum* to perform the *dilectus*, the military levy. Legion or unit lists were written, either by *scribae* or by the military tribunes themselves. One copy became the basis of individual pay and service records maintained on campaign by the quaestor and his *scribae*, while another was kept in Rome for consultation by future censors. *Lustra* performed by incoming generals ensured that field records were as accurate as possible. In combination, the various records served to support the legions to both tactical and logistical benefit while simultaneously maintaining the fair treatment of individual soldiers to prevent uprisings. It was this system which supported Rome's army as it expanded across the Mediterranean.

Mid-Republican and Imperial administration

Having estabished the nature of Mid-Republican bureaucracy, it is now possible to compare these findings with extant Imperial documents. Any similarities or differences between the two periods have further implications for this discussion. Strength reports and pay records are of particular interest. It must be noted that the surviving documents are not a representative sample of the documents generated by the Roman army during the Principate. The military documents from Vindolanda are largely concerned with supply,[24] whereas those from Dura-Europos contain much more information about manpower.[25] However, this makes the balance of their content even more significant. Between the two sites, a larger range of documentary activity is in evidence than at either one alone. More significantly here, Dura-Europos reveals that enumerating men was an important consideration in military administration, a factor somewhat masked by the predominance of logistical material at Vindolanda. That even a tiny sample of perishable documents survive from what Vegetius implies was a prodigious bureaucratic endeavour over several centuries suggests that documentation was as important as he indicates.[26]

24 *Tab. Vind.* 2.127–77.
25 *RMR passim.*
26 Vegetius, *Mil.* 2.19f.

It is unsurprising that no documents from two or three centuries earlier, in the Middle Republic, survive. The materials discovered at Vindolanda and Dura-Europos survived due to unusual soil conditions which preserved the discarded documents. No such caches have been found on Italian soil besides the documents preserved in Vesuvian ash on the Bay of Naples.[27] This lack of extant documentation is not a barrier to a comparison with the forms of paperwork proposed in this discussion. Another hurdle is that the documents from both locations deal almost exclusively with *auxilia*, not the legions, so the comparison is not quite a direct one. Nonetheless, the purpose here is to demonstrate that the forms of administration suggested are realistic. The comparison allows this. Further, even in the Republic there is an assumption that allied administration ran roughly parallel to Rome's.[28] As Imperial *auxilia* had a more integrated, permanent place within the army, it is reasonable to assume that their bureaucratic methods were not too different from those of the legions.

Some of the most famous documents from Vindolanda and Dura-Europos are strength reports. These occur in different forms, but all contain a list of a cohort's men with their rank.[29] Fink divides the papyri from Dura-Europos into daily morning reports, monthly reports and yearly *pridiana*.[30] The author follows Gilliam in being more cautious about this division;[31] the documents Fink identifies so distinctly are often very similar and fragmentary. Attribution has often been made based on similarity of form rather than an explicit mention of the type in the text. In many cases this is the only method available, but care is required nonetheless.

This is exemplified by the strength report from Vindolanda. Bowman and Thomas associate this with the *pridiana* from Egypt, but note that the form differs and that it is not dated to the end of the year.[32] This has several implications, as Bowman and Thomas mention. Firstly, it suggests that localisation of records was the norm during the Principate. While the central organisation, with the requirement for yearly reports by high officials, was uniform, the paperwork of individual legions and units may have had a more discrete form dictated by the local preferences of its creators and immediate recipients. Secondly, it suggests that there were interim strength reports generated as necessary according to the unit's needs, not just the yearly summary in the *pridiana*. Indeed, even the securest *pridiana* from Dura-Europos, *RMR* 63–4, are dated to the beginning of the Egyptian year, not the Roman. This may be another example of local idiosyncrasy, or

27 E.g., Jucundus' archive, see V:i.
28 See Introduction.
29 RMR 1–9, 47–67; Tab. Vind. 841.
30 Fink (1971) 180–217.
31 Gilliam (1962) 748, 754.
32 Bowman & Thomas (1991) 65.

it may suggest that these examples are in fact interim reports themselves. Whichever is the case, the number of strength reports and rosters indicates the importance to the Imperial armies of knowing the number and distribution of their troops, down to individual soldiers.

These strength reports reflect a concern for understanding numbers which can also be seen in the Middle Republic. Commanders took time to account for the dead after a battle when practical, keeping the legion lists as up-to-date as possible. The letters sent from the field containing the numbers and possible names of the dead indicate that such information could be obtained on a regular basis.[33] This is not direct evidence of the Imperial-type strength report, but is indicative of something similar. The expedient of accounting for the living rather than the dead is reminiscent of the roll call, which was required to create or update strength reports.[34] On the other hand, it is unlikely that strength reports from the year-long campaigns common in the fourth and earlier third centuries contained all the complexity of the Imperial records. The account from Vindolanda notes men absent in London and Coria. Such detachments were not required by the earlier, smaller armies and shorter campaigns, but there is indirect evidence of a similar system in place in the second century. In 168, reinforcements sent to Macedon who proved to be surplus to requirement were sent to garrisons.[35] As in the Vindolanda example, not all the men under a general's command were stationed in the same place. In order to keep track of these men (as this monograph argues was necessary), some kind of strength report – including the men absent in these garrisons – was required. The capability of the legions to allow for this form of garrisoning suggests that strength reports developed along with other administration during the second century.

The *pridiana* also appear to have a connection to Mid-Republican practice. If they can be described as yearly reports updating the area commander on the troops under his command, they have a similarity to the *lustra* undertaken by incoming generals. Both functioned as a yearly check of the real, rather than ideal or theoretical, state of the soldiers. It has been argued here that the *lustrum* had a written element in the form of an updated legion and unit list.[36] The yearly *lustrum* and report of the Middle Republic may have developed into the *pridianum*. Alternatively, if the *pridianum* was an interim report generated as necessary rather than a yearly institution, it nevertheless fulfils the function of giving an overview of the army's fighting strength in the same way as the 'morning' and 'monthly' reports. Reports of this type are comparable to the Mid-Republican

33 See III:ii.
34 See III:ii.
35 Livy 44.21.5–8.
36 See III:v.

general's ability to generate casualty reports. The paperwork of both periods indicates a continuing concern to know the real tactical strength of
military units.

Pay records are another area in which a remarkable similarity can be seen
between Imperial and Mid-Republican documentation. As Fink highlights,
the deductions from pay described by Polybius are exactly what can be seen
in surviving pay records on papyri.[37] *RMR* 68 is particularly striking. Fink
describes it as a record of legionary pay,[38] but Watson disagrees. Based on
the amounts cited, Watson suggests that it is the record of *deposita* showing
'the amounts standing to the men's credit' as described by Vegetius. The
records were thus produced by the clerks who assisted the standard-
bearer.[39] This document does not deal with the entire sum of pay, only
what the legionaries chose to save, but it remains a useful comparison.
The papyrus demonstrates how individual records could be kept, with
discrete deductions of different amounts made for separate men. This is
precisely the system which can be inferred from Polybius' description of
pay.[40] Once again, there is no direct evidence of documentation of this type
in the Middle Republic, but the correlation between *RMR* 68 and Polybius'
description is appealing. Indeed, it may be questioned how else the pay
records of this type were managed if not in this manner. The only major
difference to be noted is that during the Middle Republic the records would
have been under the care of the quaestor and his attendant *scribae*, not the
standard-bearer.[41]

The remarkable similarity of some of the extant Imperial documents to
those of the Middle Republic proposed here has an interesting implication
for understanding the development of Roman military administration.
Augustus is often credited with introducing complex military bureaucracy.[42] This study has demonstrated that this was not the case, but the
similarity of the documents suggests that he may have played a significant
role in its later development. Rather than introducing the documentation, it
is possible that Augustus reintroduced it. During the first century, many
traditional military structures broke down. For example, the census, so
important to military administration in the Middle Republic, was not
properly completed between 70 and 29. This failure suggests that comprehensive *tabulae iuniorum* were not produced, and so recruitment was
not undertaken in the traditional manner. This conclusion is borne out by
the evidence. First-century recruitment appears to largely have taken place

---

37 Fink (1971) 8; Polybius 6.39.12–4; *RMR* 68–73.
38 Fink (1971) 243.
39 Watson (1956) 338–9; Vegetius, *Mil.* 2.20.
40 See III:iii.
41 Note, however, that these credit amounts may also have been a feature of the second
    century BC; see Livy 40.41.8 with Boren (1983) 434.
42 See Introduction.

locally, often based on personal allegiance, in a manner more reminiscent of the *tumultus* than Polybius' *dilectus*.[43] With the acts of the warlords (including Octavian/Augustus himself), emphasised by the troubles of the Civil Wars, military organisation became increasingly about personal power and loyalty and less and less controlled by centralised, or even legal, mechanisms.

Thus, Augustus' association with military administration should be seen as reinstatement, not creation. As part of restoring the Republic, Augustus returned the census to its central position, regularised recruitment and established service terms.[44] The Imperial pay records and deployment lists suggest that, as with many things, Augustus adopted an 'if it ain't broke, don't fix it' approach. He used administrative forms that were already well established with a good track record. It is plausible that some of these records, more important for running the legion day-to-day than for macromilitary endeavours, were retained to some degree throughout the turbulent first century. Knowledge of true troop numbers and the fighting force were just as important to rogue generals in the Civil Wars as to the consuls of more stable periods, as was regular correct pay to prevent mutiny. The substantial increase in mutinies in the Late Republic indicates that the generals of this period were not as effective at preventing problems as the Mid-Republican mechanisms had been,[45] but some type of organisation was nonetheless necessary to keep their armies functioning. Either way, the evidence suggests that Augustus had an important role in the development of military bureaucracy by regularising and settling the situation – largely by reinstating the systems which preceded the first-century breakdown. If this suggestion is correct, it serves to emphasise the effectiveness of the Mid-Republican bureaucratic organisation.

All this works to demonstrate the importance of the developments in military administration undergone from the late fourth to the mid-second century. The bureaucratic forms developed in this period allowed Rome to evolve from Italian city-state to Mediterranean superpower. They facilitated many of the support mechanisms required to keep the army functioning successfully on increasingly far-flung and long-lasting campaigns. The development itself was pushed by the state's increasing demands on the abilities of its military as the scope of its interaction with neighbours – and indeed of who could be considered neighbours – extended. It is particularly noticeable in the pressures of Hannibal's invasion, but was present in some form throughout the Middle Republic. In turn, the successful evolution of bureaucratic mechanisms opened new military options to Rome. Each small change allowed Rome to look further afield. Administrative development

---

43 E.g., Appian, *BCiv.* 3.40.
44 Dio 54.26.6; *Res Gestae* 8.
45 Cf. Messer (1920) 170–3.

was not the driving force in creating Rome's empire, but it facilitated this growth by providing the necessary organisational backing for its operations. The relationship between military administration on the one hand and military success and state expansion on the other was symbiotic, each feeding the other. While it may have formed a support, developing and playing its role in empire building only because of pressures outside the symbiosis, Roman military administration is inseparable from the military itself. The bureaucratic developments of the Middle Republic remained a crucial, if more obscure, element in Roman military success by allowing the army to achieve its prescribed aims. It is the aim of this monograph to have lifted the function and functioning of this administration from its obscurity, which ancient writers accidentally created through familiarity, and to highlight both its complexity and importance in an often overlooked period of Roman bureaucratic development.

Appendix I
Men liable and available for military service

Tables A.1.1–A1.5b in Appendix I were used to calculate the approximate number of men both liable and available (that is, with a living father or brother under 60 to provide labour on a farm) to serve in the army in any given year. They have been compiled from the Coale–Demeny[2] Model Life Tables and Saller's model population simulations.[1] Level 3 West Male and Level 6 West Male have been used in conjunction with both an 'ordinary' (men aged 30, women aged 20) and 'senatorial' (men aged 25 and women aged 15) model of marriage. Tables A1.1 and A1.2 provide the male citizen population breakdowns for each Level. Tables A1.3a, A1.4a and A1.5a give the statistical likelihood of having family members and their average age in terms of these broken-down populations for each of the three models: Level 3 West Male 'ordinary' marriage, Level 3 West Male 'senatorial' marriage and Level 6 West Male 'senatorial' marriage. The totals given in Tables A1.3b, A1.4b and A1.5b were generated using probability trees.

In the model life tables, $C(x)$ is the proportion of the population within each age category. The model's proportions cannot be applied directly to the census figures; they reflect the entire male population, while the census figures count only those aged 17 plus. The total male population for each census year must be estimated first. This was done by adding together the proportions for ages 17+ in order to calculate the percentage of the population represented by the census figures. [2] For Level 3 West Male this is 60.91%, giving a total male population for 234 of 443,625 and for 164 of 553,311. [3] To these numbers, the percentages can then be applied, giving a breakdown of the men in each age category. The 10% of *proletarii* was then taken, leaving only the numbers of *assidui*. These

---

1 Coale & Demeny (1983) 107, 110; Saller (1994) 52–65 (see Tables A2.1a–A2.3a).
2 Half the percentage for 15–19 has been used. This is because the age categories represent the total between the two ages, and cannot simply be split into fifths, as the distribution is not necessarily equal across the age bracket. However, as the surrounding age brackets contain similar proportions, it is reasonable to assume a relatively even spread. On balance, half is the best way to approximate the number required.
3 All percentages given are rounded to two decimal places, and all numbers of men to the nearest person.

steps are shown in Table A1.1. The process was repeated for Level 6 West Male (Table A1.2). For Level 6 West Male the census total represented 64.51% of the whole male population, giving a total male population for 234 of 418,868 and for 164 of 584,440.

An example is provided to demonstrate the methodology: Level 3 West Male 'ordinary' marriage for ages 30–34. This was constructed using Saller's 'proportion having living kin' and 'mean age of living kin' population tables for the 30–34 age bracket.[4] Using the figure for the male population aged 30–34 (Table A1.1), the number with living wives, fathers and brothers was calculated. For the 234 population, there were 30,543 *assidui iuniores* aged 30–34. Of these, 59% had a living wife. Twenty-eight percent had a living father, with an average age of 63.8. Fifty-four percent had a living brother, with an average age of 29.5. The father can be discounted, as he was aged over 60 on average and was thus not physically able. A probability tree was used for the following calculation. Fifty-nine percent are immediately discounted due to a marriage exemption. Of the remaining 41%, 54% had a living brother. However, the model requires that 59% of these brothers were also married. Using the probability tree, this means that 9.0774% were liable and able to serve in the Level 3 West Male 'ordinary' marriage model for 30–34-year-olds using the 234 population. This gives a total of 2773.

For the 164 census figure, the number within the 30–34 age bracket is 38,095. Of these, again, 59% had a living wife, 28% had a living father, with an average age of 63.8, and 54% had a living brother, with an average age of 29.5. Following the same method, this reveals that the number of liable and able men in the Level 3 West Male 'ordinary' marriage model for 30–34-year-olds for the larger 164 population was 3458.

---

4 Saller (1994) 52–3.

Table A1.1 Level 3 West Male

Age	C(x)	234 BC (443,625)	234 BC minus 10%	164 BC (553,311)	164 BC minus 10%
0–1	3.34	14,817	13,335	18,481	16,633
1–4	9.73	43,165	38,849	53,837	48,453
5–9	10.79	47,867	43,080	59,702	53,732
10–14	10.31	45,738	41,164	57,046	51,341
15–19 (17–19)	9.82 (4.91)	43,564 (21,782)	39,208 (19,604)	54,335 (27,168)	48,902 (24,451)
20–24	9.18	40,725	36,653	50,794	45,715
25–29	8.43	37,398	33,658	46,644	41,980
30–34	7.65	33,937	30,543	42,328	38,095
35–39	6.82	30,255	27,230	37,736	33,962
40–44	5.96	26,440	23,796	32,977	29,679
45–49	5.07	22,492	20,243	28,053	25,248
50–54	4.18	18,544	16,690	23,128	20,815
55–59	3.28	14,551	13,096	18,149	16,334
60–64	2.41	10,691	9622	13,335	12,002
65–69	1.59	7054	6349	8798	7918
70–74	0.90	3993	3594	4980	4482
75–79	0.40	1775	1598	2213	1992
80–84	0.11	488	439	609	548
85+	0.02	89	80	111	100

Table A1.2 Level 6 West Male

Age	C(x)	234 BC (418,868)	234 BC minus 10%	164 BC (584,440)	164 BC minus 10%
0–1	2.72	11,393	10,254	15,897	14,307
1–4	8.66	36,274	32,647	50,613	45,552
5–9	9.92	41,552	37,397	57,975	52,178
10–14	9.58	40,128	36,115	55,989	50,390
15–19 (17–19)	9.24 (4.62)	38,703 (19,352)	34,833 (17,417)	54,002 (27,001)	48,602 (24,301)
20–24	8.77	36,735	33,062	51,255	46,130
25–29	8.22	34,431	30,988	48,041	43,237
30–34	7.63	31,960	28,764	44,593	40,134
35–39	7.00	29,321	26,389	40,911	36,820
40–44	6.31	26,431	23,788	36,878	33,190
45–49	5.57	23,331	20,998	32,553	29,298
50–54	4.79	20,064	18,058	27,995	25,196
55–59	3.96	16,587	14,928	23,144	20,830
60–64	3.08	12,901	11,611	18,001	16,201
65–70	2.19	9173	8256	12,799	11,519
70–74	1.37	5739	5165	8007	7206
75–79	0.69	2890	2601	4033	3630
80–84	0.24	1005	905	1403	1263
85–90	0.06	251	251	351	316
90–94	0.01	42	38	58	52
95+	0.00	0	0	0	0

Table A1.3a Probability of having a living relative (and the relative's average age) for Level 3 West Male 'ordinary' marriage

Age	Wife	Father	Brother
15–19	–	0.63 (49.6)	0.62 (14.8)
20–24	–	0.51 (54.3)	0.59 (19.7)
25–29	–	0.39 (59.0)	0.57 (24.7)
30–34	0.59 (21.7)	0.28 (63.8)	0.54 (29.5)
35–39	0.93 (25.1)	0.17 (68.3)	0.50 (34.5)
40–44	0.97 (29.1)	0.09 (72.7)	0.46 (39.4)

Table A1.3b Men liable and available for the levy for Level 3 West Male 'ordinary' marriage results

Age	234 BC	164 BC
15–19(17–19)	24,701(12,351)	30,808(15,404)
20–24	29,289	36,531
25–29	24,830	30,969
30–34	2772	3458
35–39	67	83
40–44	10	12
Men available 15–44 (Men available 17–44)	**81,670 (56,969)**	**101,681 (71,053)**

Table A1.4a Probability of having a living relative (and the relative's average age) for Level 3 West Male 'senatorial' marriage

Age	Wife	Father	Brother
15–19	–	0.63 (49.6)	0.62 (14.8)
20–24	–	0.51 (54.3)	0.59 (19.7)
25–29	–	0.39 (59.0)	0.57 (24.7)
30–34	0.59 (21.7)	0.28 (63.8)	0.54 (29.5)
35–39	0.93 (25.1)	0.17 (68.3)	0.50 (34.5)
40–44	0.97 (29.1)	0.09 (72.7)	0.46 (39.4)

Table A1.4b Men liable and available for the levy for Level 3 West Male 'senatorial' marriage results

Age	234 BC	164 BC
15–19 (17–19)	25,877 (12,939)	32,275 (16,138)
20–24	29,403	36,673
25–29	7675	9573
30–34	76	95
35–39	12	15
40–44	9	12
Men available 15–44 (Men available 17–44)	**63,052 (50,114)**	**78,643 (62,506)**

Table A1.5a Probability of having a living relative (and the relative's average age) for Level 6 West Male 'senatorial' marriage

Age	Wife	Father	Brother
15–19	–	0.66 (46.9)	0.59 (14.8)
20–24	–	0.54 (51.4)	0.57 (19.6)
25–29	0.59 (17.8)	0.43 (56.0)	0.54 (24.6)
30–34	0.93 (21.3)	0.32 (60.5)	0.51 (29.4)
35–39	0.97 (25.6)	0.22 (64.8)	0.48 (34.4)
40–44	0.97 (30.0)	0.13 (69.5)	0.44 (39.2)

Table A1.5b Men liable and available for the levy for Level 6 West Male 'senatorial' marriage results

Age	234 BC	164 BC
15–19 (17–19)	25,080 (12,540)	34,993 (17,497)
20–24	27,534	38,417
25–29	7858	10,964
30–34	56	78
35–39	5	7
40–44	4	6
Men available 15–44 (Men available 17–44)	60,537 (47,997)	84,465 (66,969)

Appendix II
Men over 17 years old with a living *paterfamilias*

Table A2.1a Probability of having a living relative for Level 3 West Male 'ordinary' marriage

Age	Father	Grandfather
17–19	0.60	0.10
20–24	0.51	0.04
25–29	0.39	0.01
30–34	0.28	–
35–39	0.17	–
40–44	0.09	–
45–49	0.04	–
50–54	0.01	–

Table A2.1b Probability of having a living *paterfamilias* for Level 3 West Male 'ordinary' marriage

Age	Probability
15+ (17+)	0.2861 (0.2554)
15–44 (17–44)	0.3884 (0.3565)
15–29 (17–29)	0.5338 (0.5096)

Table A2.2a Probability of having a living relative for Level 3 West Male 'senatorial' marriage

Age	Father	Grandfather
17–19	0.66	0.22
20–24	0.54	0.13
25–29	0.43	0.07
30–34	0.32	0.02
35–39	0.22	0.01
40–44	0.13	–
45–49	0.07	–
50–54	0.03	–

Table A2.2b Probability of having a living *paterfamilias* for Level 3 West Male 'senatorial' marriage

Age	Probability
15+ (17+)	0.3354 (0.3032)
15–44 (17–44)	0.4506 (0.4181)
15–29 (17–29)	0.6082 (0.5806)

Table A2.3a Probability of having a living relative for Level 6 West Male 'senatorial' marriage

Age	Father	Grandfather
17–19	0.72	0.34
20–24	0.62	0.21
25–29	0.51	0.13
30–34	0.40	0.06
35–39	0.30	0.02
40–44	0.20	0.01
45–49	0.11	–
50–54	0.05	–
55–59	0.02	–
60–64	0.01	–

Table A2.3b Probability of having a living *paterfamilias* for Level 6 West Male 'senatorial' marriage

Age	Probability
15+ (17+)	0.3826 (0.3499)
15–44 (17–44)	0.5397 (0.5070)
15–29 (17–29)	0.7098 (0.6819)

Bibliography

Unless otherwise stated, all Latin and Greek texts have been taken from the Loeb Classical Library. Translations are the author's own.

Adams, J. N., 1999. The Poets of Bu Njem: Language, Culture and Centurionate. *Journal of Roman Studies* 89, 109–34.

Adams, J. N., 2005. The Bellum Africum. In: Reinhardt, T., Lapidge, M., Adams, J. N. (eds.), *Aspects of the Language of Latin Prose*, Oxford, 73–96.

Akrigg, B., 2011. Demography and Classical Athens. In: Holleran, C., Pudsey, A. (eds.), *Demography and the Graeco-Roman World: New Insights and Approaches*, Cambridge, 37–59.

Albertoni, M., 1999. *Veiovis, Aedes (In Capitolio)*. In: Steinby, E. M. (ed.), *Lexicon Topographicum Urbis Romae, Volume Quinto: T-Z*, Rome, 99–100.

Aldrete, G., 2015. Linen Body Armour: Tests and Reconstructions at *Greek and Roman Armour Day*, London, 20th July.

Andreau, J., 1974. *Les Affaires de Monsieur Jucundus*, Rome.

Armstrong, J., 2020. Organized Chaos: *Manipuli, socii,* and the Roman Army c. 300. In: Armstrong, J., Fonda, M. P. (eds.), *Romans at War: Soliders, Citizens, and Society in the Roman Republic*, London, 76–98.

Armstrong, J., Fonda, M. P., 2020. Writing about Romans at War. In: Armstrong, J., Fonda, M. P. (eds.), *Romans at War: Soliders, Citizens, and Society in the Roman Republic*, London, 1–16.

Badian, E., 1972. *Publicans and Sinners: Private Enterprise in the Service of the Roman Republic*, Oxford.

Badian, E., 1989. The *Scribae* of the Roman Republic. *Klio* 71, 582–603.

Baird, J. A., Taylor, C., 2011. Ancient Graffiti in Context: Introduction. In: Baird, J. A., Taylor, C. (eds.), *Ancient Graffiti in Context*, London, 1–19.

Baldwin, B., 1975. *Studies in Aulus Gellius*, Lawrence.

Baronowski, D. W., 1993. Roman Military Forces in 225 B.C. (Polybius 2.23–4). *Historia* 42, 181–202.

Basanoff, V., 1950. Q. Caedicius, *Tribunus Militum* (Tradition Mythologique des Annales, 2). *Latomus* 9, 257–62.

Basanoff, V., 1951. M. Calpurnius Flamma (Tradition Mythologique des Annales, 5). *Latomus* 10, 281–4.

Beard, M., 1998. Documenting Roman Religion. In: Moatti, C. (ed.), *La Mémoire Perdue: Recherches sur l'Administration Romaine*, Paris, 3–101.

Beard, M., 2007. *The Roman Triumph*, London.

Beare, W., 1940. When Did Livius Andronicus Come to Rome? *Classical Quarterly* 34, 11–19.

Bell, M. J. V., 1965. Tactical Reform in the Roman Republican Army. *Historia* 14, 404–22.

Bergk, A., 2011. The Development of the Praetorship in the Third Century BC. In: Beck, H. *et al.* (eds.), *Consuls and Res Publica: Holding High Office in the Roman Republic*, Cambridge, 61–74.

Best, E. E., 1966. The Literate Roman Soldier. *Classical Journal* 62, 122–7.

Birley, R., 1994. *Vindolanda's Roman Record, Second Edition*, Greenhead.

Boren, H. C., 1983. Studies Relating to the *Stipendium Militum. Historia* 32, 427–460.

Botsford, G. W., 1909. *The Roman Assemblies: From Their Origin to the End of the Republic*, New York.

Bourne, F. C., 1952. The Roman Republican Census and Census Statistics. *Classical Weekly* 45, 129–35.

Bowman, A. K., 1974. Roman Military Records from Vindolanda. *Britannia* 5, 360–73.

Bowman, A. K., 1994. The Roman Imperial Army: Letters and Literacy on the Northern Frontier. In: Bowman, A. K., Woolf, G. (eds.), *Literacy and Power in the Ancient World*, Cambridge, 109–25.

Bowman, A. K., Thomas, J. D., 1983. *Vindolanda: The Latin Writing Tablets*, Gloucester.

Bowman, A. K., Thomas, J. D., 1991. A Military Strength Report from Vindolanda. *Journal of Roman Studies* 81, 62–73.

Boyd, C. E., 1915. *Public Libraries and Literary Culture in Ancient Rome*, Chicago.

Breeze, D. J., 1971. Pay Grades and Ranks below the Centuriate. *Journal of Roman Studies* 61, 130–35.

Brennan, T. C., 1996. Triumphs on Monte Albano. In: Wallace, R. W., Harris, E. H. (eds.), *Transitions to Empire: Essays in Greco-Roman History 360–146 B.C., in Honor of E. Badian*, London, 315–37.

Brennan, T. C., 2004. Power and Process under the Republican 'Constitution'. In: Flowers, H. I.(ed.), *The Cambridge Companion to the Roman Republic*, Cambridge, 31–65.

Brink, C. O., Walbank, F. W., 1954. The Construction of the Sixth Book of Polybius *Class. Q.* 4, 97–122.

Briscoe, J., 1971. The First Decade. In: Dorey, T. A. (ed.), *Livy*, Edinburgh, 1–20.

Briscoe, J., 1973. *A Commentary on Livy Books XXXI–XXXIII*, Oxford.

Briscoe, J., 1981. *A Commentary on Livy Books XXXIV–XXXVII*, Oxford.

Briscoe, J., 2008. *A Commentary on Livy Books 38–40*, Oxford.

Briscoe, J., 2009. Livy's Sources and Methods of Composition in Books 31–33. In: Chaplin, J. D., Kraus, C. S. (eds.), *Livy*, Oxford, 461–75.

Briscoe, J., 2012. *A Commentary on Livy Books 41–45*, Oxford.

Broughton, T. R. S., 1951. *The Magistrates of the Roman Republic: Volume I, 509 BC–100 BC*, Oxford.

Brunt, P. A., 1950. Pay and Superannuation in the Roman Army. *Papers of the British School at Rome* 18, 50–71.

Brunt, P. A., 1966. The 'Fiscus' and Its Development. *Journal of Roman Studies* 56, 75–91.

Brunt, P. A., 1971. *Italian Manpower, 225B.C.–A.D.14*, Oxford.

Bucher, G. S., 1987. The *Annales Maximi* in the Light of Roman Methods of Keeping Records. *American Journal of Ancient History* 12, 2–63.

Buckland, W. W., 1908. *The Roman Law of Slavery: The Condition of the Slave in Private Law from Augustus to Justinian*, Cambridge.

Buettner-Wobst, T. (ed.), 1889. *Polybii Historiae, Vol. I, Libri IV–VIII*, Stuttgart.

Burck, E., 1971. The Third Decade. In: Dorey, T. A. (ed.), *Livy*, Edinburgh, 21–46.

Calboli, G., 1996. Die Episode des Tribunen Q. Caedicius (Cato, Orig. Frg. 7–43 Peter). *MAIA* 48, 1–32.

Campbell, B., 1987. Teach Yourself to Be a General. *Journal of Roman Studies* 77, 13–29.

Cavaignac, E., 1914. Le Texte de Polybe VI, 19, 2 et la Durée du Service Militaire a Rome. *Revue de Philologie* 38, 76–80.

Champion, C. B., 2004. *Cultural Politics in Polybius' Histories*, London.

Chaplin, J. D., 2000. *Livy's Exemplary History*, Oxford.

Chaplin, J. D., Kraus, C. S. (eds.), 2009. *Livy*, Oxford.

Chrissanthos, S. G., 1997. Scipio and the Mutiny at Sucro, 206 B.C. *Historia* 46, 172–84.

Christ, M. R., 2007. The Evolution of the *Eisphora* in Classical Athens. *Classical Quarterly* 57.1, 53–69.

Coale, A. J., Demeny, P., 1983. *Regional Model Life Tables and Stable Populations, Second Edition*, London.

Coarelli, F., 1993. *Atrium Libertatis*. In: Steinby, E. M. (ed.), *Lexicon Topographicum Urbis Romae, Volume Primo: A-C*, Rome, 133–35.

Coarelli, F., 1997. *Il Campo Marzio: Dalee origini alla fine della Repubblica*, Rome.

Coarelli, F., 1999. *Saturnus, Aedes*. In: Steinby, E. M. (ed.), *Lexicon Topographicum Urbis Romae, Volume Quarto: P-S*, Rome, 234–6.

Coarelli, F., 2007. *Rome and Environs: An Archaeological Guide*, London.

Coarelli, F., 2010. *Substructio et Tabularium. Papers of the British School in Rome* 78, 107–32.

Corbier, M., 1974. *L'Aerarium Saturni et L'Aerarium Militare*, Rome.

Corcoran, S., 2014. State Correspondence in the Roman Empire: Imperial Communication from Augustus to Justinian. In: Radner, K. (ed.), *State Correspondence in the Ancient World: From New Kingdom Egypt to the Roman Empire*, Oxford, 172–210.

Cornell, T. J., 1991. The Tyranny of the Evidence: A Discussion of the Possible Uses of Literacy in Etruria and Latium in the Archaic Age. In: Humphrey, J. (ed.), *Literacy in the Roman World*, Ann Arbor, 7–34.

Cornell, T. J., 1995. *The Beginnings of Rome: Italy and Rome from the Bronze Age to the Punic Wars (c. 1000–264 BC)*, London.

Cornell, T. J. *et al.*, 2013a. *The Fragments of the Roman Historians: Volume 1, Introduction*, Oxford.

Cornell, T. J. *et al.*, 2013b. *The Fragments of the Roman Historians: Volume 2, Text and Translation*, Oxford.

Cornell, T. J. *et al.*, 2013c. *The Fragments of the Roman Historians: Volume 3, Commentary*, Oxford.

Coudry, M., 1994. Sénatus-Consultes et *Acta Senatus*: Rédaction, Conservation et Archivage des Documents Émanand de Sénat, De l'époque de César à celle des

Sévères. In: Demougin, S. (ed.), *La Mémoire Perdue: A la recherche des archives oubliées, publiques et privies de la Roma antique*, Paris, 65–102.

Crawford, M. H., 1985. *Coinage and Money under the Roman Republic: Italy and the Mediterranean Economy*, London.

Crawford, M. H. *et al.* (ed.), 1996. *Roman Statutes, Volume I*, London.

Culham, P., 1989. Archives and Alternatives in Republican Rome. *Classical Philology* 84, 100–15.

Daly, G., 2002. *Cannae: The Experience of Battle in the Second Punic War*, London.

Davies, R. W., 1967. *Ratio* and *Opinio* in Roman Military Documents. *Historia* 16, 115–18.

Degrassi, A., 1947. *Fasti Consulares et Triumphales. Inscriptiones Italiae, Vol. 13*, Rome.

De Ligt, L., 2004. Poverty and Demography: The Case of the Gracchan Land Reforms. *Mnemosyne* 57, 725–57.

De Ligt, L., 2007. Roman Manpower and Recruitment During the Middle Republic. In: Erdkamp, P. (ed.), *A Companion to the Roman Army*, Oxford, 114–31.

De Sanctis, G., 1917. *Storia dei Romani, Volume III: L'età della guerre Puniche, Parte II*, Torino.

Develin, R., 1978. Tradition and the Development of Triumphal Regulations in Rome. *Klio* 60, 429–38.

Dittman, M., 1935. The Development of Historiography among the Romans. *Classical Journal* 30, 287–96.

Dix, T. K., 1994. 'Public Libraries' in Ancient Rome: Ideology and Reality. *Libraries and Culture* 29, 282–96.

Dobson, M., 2008. *The Army of the Roman Republic: The Second Century BC, Polybius and the Camps at Numantia, Spain*, Oxford.

Drogula, F. K., 2015. *Commanders and Command in the Roman Republic and Early Empire*, Chapel Hill.

Duncan-Jones, R., 1982. *The Economy of the Roman Empire: Quantitative Studies, Second Edition*, Cambridge.

Eckstein, A. M., 1987. *Senate and General: Individual Decision Making and Roman Foreign Relations, 264–194 B.C.*, London.

Eckstein, A. M., 1995. *Moral Vision in the Histories of Polybius*, London.

Eckstein, A. M., 1997. *Physis* and *Nomos*: Polybius, the Romans and Cato the Elder. In: Cartledge, P., Garnsey, P., Gruen, E. (eds.), *Hellenistic Constructs: Essays in Culture, History and Historiography*, London, 175–99.

Edmondson, J., 1993. *Instrumenta Imperii*: Law and Imperilaism in Republican Rome. In: Halpern, B., Hobson, D. W. (eds.), *Law, Politics and Society in the Ancient Mediterranean World*, Sheffield, 156–92.

Erdkamp, P., 1995. The Corn Supply of the Roman Armies during the Third and Second Centuries B.C. *Historia* 44, 168–91.

Erdkamp, P., 1998. *Hunger and the Sword: Warfare and Food Supply in Roman Republican Wars (264–30 B.C.)*, Amsterdam.

Erdkamp, P., 2002. Introduction. In: Erdkamp, P. (ed.), *The Roman Army and the Economy*, Amsterdam, 5–16.

Erdkamp, P., 2006a. Late-Annalistic Battle Scenes in Livy (Books 21–44). *Mnemosyne* 59, 525–63.

Erdkamp, P., 2006b. Valerius Antias and Livy's Casualty Reports. *Studies in Latin Literature and Roman History* 13, 166–82.

Erdkamp, P., 2007. War and State Formation in the Roman Republic. In: Erdkamp, P. (ed.), *A Companion to the Roman Army*, Oxford, 96–113.

Erdkamp, P., 2009. Polybius, the Ebro Treaty and the Gallic Invasion of 225 B.C.E. *Classical Philology* 104, 495–510.

Erskine, A., 2012. Polybius among the Romans: Life in the Cyclops' Cave. In: Smith, C., Yarrow, L. M. (eds.), *Imperialism, Cultural Politics, and Polybius*, Oxford, 17–32.

Erskine, A., 2013. How to Rule the World: Polybius Book 6 Reconsidered. In: Gibson, B., Harrison, T. (eds.), *Polybius and His World: Essays in Memory of F. W. Walbank*, Oxford, 231–46.

Feig Vishnia, R., 2002. The Shadow Army: The *Lixae* and the Roman Legions. *Zeitschrift für Papyrologie und Epigraphik* 139, 265–72.

Feldherr, A., 1998. *Spectacle and Society in Livy's* History, London.

Fink, R. O., 1971. *Roman Military Records on Papyrus*, London.

Flower, H. I., 2010. *Roman Republics*, Oxford.

France, J., 2006. *Tributum et Stipendium*: La Politique Fiscal de l'Empereur Romain. *Revue Historique de Droit Français et Étranger* 84, 1–17.

Frank, T., 1924. Roman Census Statistics from 225 to 28 B.C. *Classical Philology* 24, 329–41.

Frank, T., 1930. Roman Census Statistics from 508 to 225 B.C. *American Journal of Philology* 51, 313–24.

Frank, T., 1932. The Public Finances of Rome 200–157 B.C. *American Journal of Philology* 53, 1–20.

Frank, T., 1933. *An Economic Survey of Ancient Rome, Volume I: Rome and Italy of the Republic*, Baltimore.

Frederiksen, M. W., 1965. The Republican Municipal Laws: Errors and Drafts. *Journal of Roman Studies* 55, 183–98.

Frier, B. W., 1975. Licinius Macer and the *Consules Suffecti* of 444 B.C. *Transactions of the American Philological Association* 105, 79–97.

Gabba, E., 1951. Ricerche sull' esercito professionale romano da Mario ad Augusto. *Athenaeum* 29, 171–272.

Gabba, E., 1976. *Republican Rome, the Army and the Allies* Cuff, P.J. (trans.), Berkeley.

Garnsey, P., Saller, R., 1987. *The Roman Empire: Economy, Society and Culture*, London.

Gilliam, J. F., 1962. The Moesian *Pridianum*. *Hommages à Albert Grenier, Latomus* 58, 747–56.

Gilliver, C., 1999. *The Roman Art of War*, Stroud.

Gilliver, K., 2001. Feeding an Army. *The Classical Review* 51, 344–7.

Goldberg, S. M., 1986. *Understanding Terence*, Princeton.

Golden, G. K., 2013. *Crisis Management During the Roman Republic: The Role of Political Institutions in Emergencies*, Cambridge.

Greep, S. J., 1987. Lead Sling-Shot from Windridge Farm, St Albans and the Use of the Sling by the Roman Army in Britain. *Britannia* 18, 183–200.

Gudeman, A., 1894. Literary Frauds among the Romans. *Transactions of the American Philological Association* 25, 140–64.

Gurd, S., 2010. Verres and the Scene of Rewriting. *Phoenix* 64, 80–101.

Haines, C. R. (ed.), 1982. *The Correspondence of Marcus Cornelius Fronto with Marcus Aurelius Antoninus, Lucius Verus, Antoninus Pius, and Various Friends*, London.

Hamp, E. P., 1986. Notes on Latin Noun Formation. *Rheinisches Museum für Philologie* 129, 362.

Hansen, M. H., 2006. *The Shotgun Method: The Demography of the Ancient Greek City-State Culture*, Columbia.

Hanson, A. E., 1991. Ancient Illiteracy. In: Humphrey, J. H. (ed.), *Literacy in the Roman World*, Ann Arbor, 159–98.

Hardie, A., 2007. Juno, Hercules, and the Muses at Rome. *American Journal of Philology* 128, 551–92.

Hardy, E. G., 1912. *Three Spanish Charters and Other Documents: Translated with Introductions and Notes*, Oxford.

Hardy, E. G., 1914. The Table of Heraclea and the *Lex Iulia Municipalis*. *Journal of Roman Studies* 4, 65–110.

Harris, W. V., 1971. *Rome in Etruria and Umbria*, Oxford.

Harris, W. V., 1976. The Development of the Quaestorship, 267–81 B.C. *Classical Quarterly* 26, 92–106.

Harris, W. V., 1985. *War and Imperialism in Republican Rome, 327–70 B.C., Second Edition*, Oxford.

Harris, W. V., 1989. *Ancient Literacy*, London.

Helm, M., 2020. Poor Man's War – Rich Man's Fight: Military Integration in Republican Rome. In: Armstrong, J., Fonda, M. P. (eds.), *Romans at War: Soliders, Citizens, and Society in the Roman Republic*, London, 99–115.

Herrman, L., 1946. Ennius et les livres de Numa. *Latomus* 5, 87–90.

Hill, H., 1939. Census Equester. *American Journal of Philology* 60, 357–62.

Hill, H., 1946. The History of *Pignoriscapio*. *American Journal of Philology* 46 (1), 60–66.

Hin, S., 2008. Counting Romans. In: de Ligt, L., Northwood, S. (eds.), *People, Land, and Politics: Demographic Developments and the Transformation of Roman Italy 300 BC–AD 14*, Leiden, 187–238.

Hin, S., 2013. *The Demography of Roman Italy: Population Dynamics in an Ancient Conquest Society 201 BCE–14 CE*, Cambridge.

Holford-Strevens, L., 1988. *Aulus Gellius*, London.

Holleran, C., Pudsey, A., 2011. Introduction: Studies in Ancient Historical Demography. In: Holleran, C., Pudsey, A. (eds.), *Demography and the Graeco-Roman World: New Insights and Approaches*, Cambridge, 1–13.

Hopkins, K., 1978. *Conquerors and Slaves*, Cambridge.

Housten, G. W., 2002. The Slave and Freedmen Personnel of Public Libraries in Ancient Rome. *Transactions of the American Philological Association* 132, 139–76.

Howard, A. A., 1906. Valerius Antias and Livy. *Harvard Studies in Classical Philology* 17, 161–82.

Huntley, K. V., 2011. Identifying Children's Graffiti in Roman Campania: A Developmental Psychological Approach. In: Baird, J. A., Taylor, C. (eds.), *Ancient Graffiti*, London, 69–89.

Isayev, E., 2017. *Migration, Mobility and Place in Ancient Italy*, Cambridge.

Jacobsthal, P., 1943. On Livy XXXVI, 40 (Boiian Silver). *American Journal of Archaeology* 47, 306–12.

Jaeger, M., 1997. *Livy's Written Rome*, Ann Arbor.

Jones, A. H. M., 1949. The Roman Civil Service (Clerical and Sub-Clerical Grades). *Journal of Roman Studies* 39, 38–55.

Jongman, W., 1988. *The Economy and Society of Pompeii*, Amsterdam.

Kay, P., 2014. *Rome's Economic Revolution*, Oxford.

Keppie, L., 1984. *The Making of the Roman Army: From Republic to Empire*, Manchester.

Klotz, A., 1937. Diodors Römische Annalen. *Rheinisches Museum für Philologie* 86, 206–24.

Koon, S., 2010. *Infantry Combat in Livy's Battle Narratives*, Oxford.

Krentz, P., 1985. Casualties in Hoplite Battles. *Greek, Roman and Byzantine Studies* 26, 13–20.

Lange, C. H., 2016. *Triumphs in the Age of Civil War: The Late Republic and the Adaptability of Triumphal Tradition*, London.

Langslow, D., 2013. Archaic Inscriptions and Greek and Roman Authors. In: Liddel, P., Low, P. (eds.), *Inscriptions and Their Uses in Greek and Latin Literature*, Oxford, 167–96.

Laroche, R. A., 1988. Valerius Antias: Livy's Source for the Number of Military Standards Captured in Battle in Books XX–XLV. *Latomus* 47, 758–71.

Latte, K., 1936. The Origin of the Roman Quaestorship. *Transactions and Proceedings of the American Philological Association* 67, 24–33.

Lazenby, J. F., 1978. *Hannibal's War: A Military History of the Second Punic War*, Warminster.

Le Bohec, Y., 2015. Roman Wars and Armies in Livy. In: Mineo, B. (ed.), *A Companion to Livy*, Hoboken.

Leeman, A. D., 1963. *Orationis Ratio: The Stylistic Theories and Practice of the Roman Orators, Historians and Philosophers, Volume I*, Amsterdam.

Lendon, J. E., 1999. The Rhetoric of Combat: Greek Military Theory and Roman Culture in Julius Caesar's Battle Descriptions. *Classical Antiquity* 18, 237–329.

Lendon, J. E., 2005. *Soldiers and Ghosts: A History of Battle in Antiquity*, New Haven.

Levene, D. S., 2010. *Livy on the Hannibalic War*, Oxford.

Lintott, A., 1993. Imperium Romanum: *Politics and Administration*, London.

Lintott, A., 1999. *The Constitution of the Roman Republic*, Oxford.

Lo Cascio, E., 1990. Le *professions* della *Tabula Heracleensis* e la procedure del *census* in età Caesariana. *Athenaeum* 78, 287–318.

Lo Cascio, E., 1999. The Population of Italy in Town and Country. In: Bintliff, J., Sbonias, K. (eds.), *Reconstructing Past Population Trends in Mediterranean Europe (3000 BC–AD 1800)*, Oxford, 161–71.

Lo Cascio, E., 2001. Recruitment and the Size of the Roman Population from the Third to the First Century BCE. In: Scheidel, W. (ed.), *Debating Roman Demography*, Leiden, 111–38.

Luce, T. J., 1977. *Livy: The Composition of his History*, Princeton.

Lugli, G., 1947. *Monumenti minori del foro Romano*, Rome.

MacMullen, R., 1980. How Many Romans Voted? *Athenaeum* 58, 454–7.

Marchetti, P., 1976. *Histoire Économique et Monétaire de la Deuxième Guerre Punique*, Brussels.

Marchetti, P., 1977. A Propos du *Tributum* Romain: Impôt de Quotité ou de Répartition. In: Chastagnol, H., Nicolet, C., van Effenterre, H. (eds.) *Armées et Fiscalité dans le Monde Antique*, Paris, 107–33.

Marquardt, J., 1891. *De L'Organisation Militaire Chez les Romains* Brissaud, M. (trans.), Paris.

Marsden, E. W., 1974. Polybius as a Military Historian. In: Gabba, E. (ed.), *Polybe: Neuf Exposés Suivis De Discussions*, Genève, 267–95.

Massa-Pairault, F.-H., 2001. Relations d'Appius Claudius Caecus avec l'Etrurie et la Campanie. In: Briquel, D., Thuillier, J.-P. (eds.) *Le Censeur et les Samnites: sur Tite Live, livre IX*, Paris, 97–116.

Mattingly, H. B., 1957. The Date of Livius Andronicus. *Classical Quarterly* 7, 159–63.

McGing, B., 2010. *Polybius' Histories*, Oxford.

Meadows, A., Williams, J., 2001. Moneta and the Monuments: Coinage and Politics in Republican Rome. *Journal of Roman Studies* 91, 27–49.

Messer, W. S., 1920. Mutiny in the Roman Army. The Republic. *Classical Philology* 15, 158–75.

Meyer, E. A., 2004. *Legitimacy and Law in the Roman World: Tabulae in Roman Belief and Practice*, Cambridge.

Middleton, P., 1983. The Roman Army and Long Distance Trade. In: Garnsey, P., Whittaker, C. R. (eds.), *Trade and Famine in Classical Antiquity*, Cambridge, 75–83.

Miles, G. B., 1995. *Livy: Reconstructing Early Rome*, London.

Millar, F., 1964. The *Aerarium* and Its Officials under the Empire. *Journal of Roman Studies* 54, 33–40.

Mohr Mersing, K., 2007. The War-tax (*Tributum*) of the Roman Republic: A Reconsideration. *Classica et Mediaevalia* 58, 215–35.

Momigliano, A., 1975. *Alien Wisdom: The Limits of Hellenisation*, Cambridge.

Momigliano, A., 1977. The Historian's Skin In:. *Essays in Ancient and Modern Historiography*, Oxford, 67–77.

Mommsen, T., 1859. *Die Römische Chronologie bis auf Caesar*, 2nd Edition, Berlin.

Mommsen, T., 1893. *Le Droit Public Romain, Tome Premier* Girard, P.F. (trans.), Paris.

Mommsen, T., 1894. *Le Droit Public Romain, Tome Quatrième* Girard, P.F. (trans.), Paris.

Moore, J. M., 1965. *The Manuscript Tradition of Polybius*, Cambridge.

Mouritsen, H., 2001. *Plebs and Politics in the Late Republic*, Cambridge.

Mouritsen, H., 2011. *The Freedman in the Roman World*, Cambridge.

Ñaco del Hoyo, T., 2011. Roman Economy, Finance and Politics in the Second Punic War. In: Hoyos, D. (ed.), *A Companion to the Punic Wars*, Oxford.

Ñaco del Hoyo, T., 2019. Rethinking *Stipendiarius* as Terminology of the Roman Republic: Political and Military Dimensions. *Museum Helveticum* 76 (1), 70–87.

Newell, C., 1988. *Methods and Models in Demography*, Chichester.

Nicolet, C., 1976. Tributum: *Recherches sur la Fiscalité Directe sous la Republique Romaine*, Bonn.

Nicolet, C., 1980. *The World of the Citizen in Republican Rome* Falla, P.S. (trans.), London.

Nicolet, C., 2000. *Censeurs et Publicains: Économie et Fiscalité dans la Rome Antique*, Paris.

Northwood, S., 2008. Census and *Tributum*. In: de Ligt, L., Northwood, S. (eds.), *People, Land and Politics: Demographic Developments and the Transformation of Roman Italy 300 BC–AD 14*, Leiden, 257–72.

Oakley, S. P., 1997. *A Commentary on Livy Books VI–X, Volume I: Introduction and Book VI*, Oxford.

Oakley, S. P., 1998. *A Commentary on Livy Books VI–X, Volume II: Books VII–VIII*, Oxford.

Oakley, S. P., 2005. *A Commentary on Livy Books VI–X, Volume III: Book IX*, Oxford.

Oakley, S. P., 2009. Livy and his Sources. In: Chaplin, J.D., Kraus, C.S. (eds.), *Livy*, Oxford, 439–60.

Ogilvie, R. M., 1958. Livy, Licinius Macer and the *Libri Lintei. Journal of Roman Studies* 48, 40–46.

Ogilvie, R. M., 1961. *Lustrum Condere. Journal of Roman Studies* 51, 31–39.

Ogilvie, R. M., 1965. *A Commentary on Livy, Books 1–5*, Oxford.

Oliver, J. H., Palmer, R. E. A., 1954. Text of the *Tabula Hebana. American Journal of Philology* 75, 225–49.

Östenberg, I., 2009. *Staging the World: Spoils, Captives, and Representations in the Roman Triumphal Procession*, Oxford.

Otto, W. F., 1916. *Lustrum. Rheinisches Museum für Philologie* 71, 17–40.

Pallottino, M., 1955. *The Etruscans* Cremona, J. (trans.), London.

Parkin, T. G., 1992. *Demography and Roman Society*, London.

Patterson, M. L., 1942. Rome's Choice of Magistrates during the Hannibalic War. *Transactions and Proceedings of the American Philological Association* 73, 319–40.

Pearson, E. H., 2012. *Livy and the Mutiny of 342 BC: A Political and Socio-Economic Case Study of the Fourth Century*, Unpublished MA Thesis, Manchester.

Pearson, E. H., 2018. Political Space. In: Holleran, C., Claridge, A. (eds.), *A Companion to the City of Rome*, Chichester, 559–79.

Pelikan Pittenger, M. R., 2008. *Contested Triumphs: Politics, Pageantry, and Performance in Livy's Republican Rome*, London.

Pelling, C., 2007. The Greek Historians of Rome. In: Marincola, J. (ed.), *A Companion to Greek and Roman Historiography, Volume I*, Oxford, 244–58.

Perjes, G., 1970. Army Provisioning, Logistics and Strategy in the Second Half of the 17th Century. *Acta Historia Academiae Scientiarum Hungariae* 16, 1–51.

Pfister, R., Belinger, L., 1945. *The Excavations at Dura-Eurpos, Final Report IV. Part II: The Textiles*, London.

Phang, S. E., 2007. Military Documents, Languages, and Literacy. In: Erdkamp, P. (ed.), *A Companion to the Roman Army*, Oxford, 286–305.

Pina Polo, F., Díaz Fernández, A., 2019. *The Quaestorship in the Roman Republic*, Berlin.

Platner, S. B., Ashby, T., 1929. *A Topographical Dictionary of Ancient Rome*, London.

Posner, E., 1972. *Archives in the Ancient World*, Cambridge.

Potter, D., 2004. The Roman Army and Navy. In: Flower, H.I. (ed.), *The Cambridge Companion to the Roman Republic*, Cambridge, 66–88.

Prag, J. R. W., 2014. The Quaestorship in the Third and Second Centuries BC. In: Doubouloz, J., Pittia, S., Sabatini, G. (eds.), *L'Imperium Romanum en Perspective: Les Savoirs d'Empire dans le République Romaine et leur Héritage dans l'Europe Médiévale et Moderne*, Besançon, 193–209.

Purcell, N., 1983. The *Apparitores*: A Study in Social Mobility. *Papers of the British School at Rome* 51, 125–73.

Purcell, N., 1993. *Atrium Libertatis*. *Papers of the British School at Rome* 61, 125–55.

Purcell, N., 2001. The *Ordo Scribarum*: A Study in the Loss of Memory. *Mélanges de l'Ecole Française de Rome, Antiquité* 113, 633–74.

Rathbone, D., 1993. The Census Qualification of the *Assidui* and the *Prima Classis*. In: Sancisi-Weerdenburg, H., Van Der Spek, R. J., Teitler, H. C., Wallinga, H. T. (eds.), De Agricultura: In Memoriam *Pieter Willem De Neeve (1945–1990)*, Amsterdam, 121–52.

Rathbone, D., 2007. Military Finance and Supply. In: Sabin, P. *et al.* (eds.), *The Cambridge History of Greek and Roman Warfare. Volume I: Greece, the Hellenistic World and the Rise of Rome*, Cambridge, 158–76.

Rawson, E., 1971. The Literary Sources for the Pre-Marian Army. *Papers of the British School at Rome* 39, 13–31.

Rawson, E., 1985. *Intellectual Life in the Late Roman Republic*, London.

Reid, J. S., 1915. The So-Called *Lex Iulia Municipalis*. *Journal of Roman Studies* 5, 207–48.

Rich, J. W., 1983. The Supposed Roman Manpower Shortage of the Later Second Century B.C. *Historia* 32, 287–31.

Rich, J. W., 2005. Valerius Antias and the Construction of the Roman Past. *Bulletin of the Institute of Classical Studies* 48, 137–61.

Rich J. W., 2013. *Annales* and *Annales Maximi*: The Origins of the Roman Annalistic Tradition. Unpublished at Omnium annalium monumenta: *Historical Evidence and Historical Writing in Republican Rome*, Rome, 31st October.

Richardson, J. S., 1986. *Hispaniae: Spain and the Development of Roman Imperialism, 218–82 BC*, Cambridge.

Richardson, J. S., 2000. *Appian: Wars of the Romans in Iberia, with an Introduction, Translation and Commentary*, Warminster.

Richardson, J. S., 2014. Firsts in the Historians of Rome. *Historia* 63.1, 17–37.

Richardson, L., 1980. The Approach to the Temple of Saturn in Rome. *American Journal of Archaeology* 84, 51–62.

Richardson, L., 1992. *A New Topographical Dictionary of Ancient Rome*, London.

Rickman, G., 1980. *The Corn Supply of Ancient Rome*, Oxford.

Roberts, L. G., 1918. The Gallic Fire and the Roman Archives. *Memoirs of the American Academy in Rome* 2, 55–65.

Rosenberger, V., 2003. The Gallic Disaster. *Classical World* 96, 365–373.

Rosenstein, N., 2002. Marriage and Manpower in the Hannibalic War: *Assidui*, *Proletarii* and Livy 24.18.7–8. *Historia* 51, 163–91.

Rosenstein, N., 2004. *Rome at War: Farms, Families, and Death in the Middle Republic*, London.

Rosenstein, N., 2011. War, Wealth and Consuls. In: Beck, H. *et al.* (eds.), *Consuls and* Res Publica: *Holding High Office in the Roman Republic*, Cambridge, 133–58.

Rosenstein, N., 2012. *Rome and the Mediterranean 290 to 146 BC: The Imperial Republic*, Edinburgh.

Rosenstein, N., 2016a. *Bellum se ipsum alet?*: Financing Mid-Republican Imperialism. In: Beck, H., Jehne, M., Serrati, J. (eds.), *Money and Power in the Roman Republic*, Brussels, 114–30.

Rosenstein, N., 2016b. *Tributum* in the Middle Republic. In: Armstrong, J. (ed.), Circum Mare: *Themes in Ancient Warfare*, Leiden, 80–97.

Rosenstein, N. (forthcoming). Paying for Conquest over the 'Long Fourth Century'. In: Peralta, D. P., Bernard, S., Mignone, L. (eds.), *Republican Rome in the Long Fourth Century*, Cambridge.

Rosivach, V. J., 1983. Mars, the Lustral God. *Latomus* 42, 509–21.

Roth, J., 1994. The Size and Organisation of the Roman Imperial Legion. *Historia* 43, 346–62.

Roth, J., 1999. *The Logistics of the Roman Army at War (246 B.C.–A.D. 235)*, Leiden.

Rowan, C., 2013. The Profits of War and Cultural Capital: Silver and Society in Republican Rome. *Historia* 62 (3), 361–86.

Ryan, F. X., 1998. *Rank and Participation in the Republican Senate*, Stuttgart.

Sabin, P., 1996. The Mechanics of Battle in the Second Punic War. In: Cornell, T., Rankov, B., Sabin, P. (eds.), *The Second Punic War: A Reappraisal*, London, 59–80.

Sage, M. M., 2008. *The Republican Roman Army: A Sourcebook*, London.

Sallares, R., 2002. *Malaria and Rome: A History of Malaria in Ancient Italy*, Oxford.

Saller, R. P., 1994. *Patriarchy, Property and Death in the Roman Family*, Cambridge.

Scheidel, W., 1996. Finances, Figures and Fiction. *Classical Quarterly* 46, 222–38.

Scheidel, W., 2004. Human Mobility in Roman Italy, I: The Free Population. *Journal of Roman Studies* 94, 1–26.

Scheidel, W., 2009. The Demographic Background. In: Hübner, S. R., Ratzan, D. M. (eds.), *Growing Up Fatherless in Antiquity*, Cambridge, 31–40.

Serrati, J., 2007. Warfare and the State. In: Sabin, P. *et al.* (eds.), *The Cambridge History of Greek and Roman Warfare. Volume I: Greece, the Hellenistic World and the Rise of Rome*, Cambridge, 461–97.

Shatzman, I., 1972. The Roman General's Authority over Booty. *Historia* 21 (2), 117–205.

Sherk, R. K., 1969. *Roman Documents from the Greek East: Senatus Consulta and Epistulae to the Age of Augustus*, Baltimore.

Shochat, Y., 1980. *Recruitment and the Programme of Tiberius Gracchus*, Bruxelles.

Sommella Mura, A., 1981. Il Tabularium: Pogetto di Consolidamento e Restauro. *Archeologia Laziale* 4, 126–31.

Soraci, C., 2010. Riflessioni Storico-comparative sul Terminus *Stipendiarius*. In: Cataudella, M. R., Greco, A., Marriotta, G. (eds.), *Strumenti tecniche della riscossione dei tribute nel mondo antico. Atti del Convegno Nazionale Firenze 6–7 Dicembre 2007*, Padua.

Southern, P., 2007. *The Roman Army: A Social and Institutional History*, Oxford.

Speidel, M. A., 1992. Roman Army Pay Scales. *Journal of Roman Studies* 82, 87–106.

Stevenson, A. J., 2004. Gellius and the Roman Antiquarian Tradition. In: Holford-Strevens, L., Vardi, A. (eds.), *The Worlds of Aulus Gellius*, Oxford, 118–55.

Still, K. G., 2014. *Introduction to Crowd Science*, London.

Still, K. G., 2019. Visualising Crowd Density. at www.gkstill.com/Support/crowd-density/CrowdDensity-2.html, last accessed 15 August 2020.

Sumner, G. V., 1970. The Legion and the Centuriate Organisation. *Journal of Roman Studies* 60, 67–78.

Suolahti, J., 1955. *The Junior Officers of the Roman Army in the Republican Period: A study on Social Structure*, Helsinki.

Suolahti, J., 1963. *The Roman Censors: A Study on Social Structure*, Helsinki.

Sutherland, C. H. V., 1945. *Aerarium* and *Fiscus* during the Early Empire. *American Journal of Philology* 66, 151–70.

Syme, R., 1959. The Lower Danube under Trajan. *Journal of Roman Studies* 49, 26–33.

Tan, J., 2013. Booty and the Roman Assembly in 264 B.C. *Historia* 62.4, 417–9.

Tan, J., 2015. The Roman Republic. In: Monson, A., Scheidel, W. (eds.), *Fiscal Regimes and the Political Economy of Premodern States*, Cambridge, 208–28.

Tan, J., 2017. *Power and Public Finance at Rome, 264–49 BCE*, Oxford.

Tan, J., 2020. The *Dilectus-Tributum* System and the Settlement of Fourth Century Italy. In: Armstrong, J., Fonda, M. P. (eds.), *Romans at War: Soldiers, Citizens and Society in the Roman Republic*, London, 52–75.

Tan, J. (forthcoming). The Long Shadow of *Tributum* in the Long Fourth Century. In: Peralta, D. P., Bernard, S., Mignone, L. (eds.), *Republican Rome in the Long Fourth Century*, Cambridge.

Taylor, H. G., 2012. Fragments of Linen from Masada, Israel – the Remnants of Pteryges? – and Related Finds in Weft- and Warp-Twining Including Several Slings. In: Nosch, M.-L. (ed.), *Wearing the Cloak: Dressing the Soldier in Roman Times*, Oxford, 56–84.

Taylor, L. R., 1946. The Date of the Capitoline Fasti. *Classical Philology* 41, 1–11.

Taylor, L. R., 1957. The Centuriate Assembly before and after the Reform. *American Journal of Philology* 78, 337–54.

Taylor, L. R., 1960. *The Voting Districts of the Roman Republic: The Thirty-Five Urban and Rural Tribes*, Rome.

Taylor, L. R., 1966. *Roman Voting Assemblies: From the Hannibalic War to the Dictatorship of Caesar*, Birmingham.

Taylor, M. J., 2014. Roman Infantry Tactics in the Mid-Republic: A Reassessment. *Historia* 63, 301–22.

Taylor, M. J., 2017. State Finance in the Middle Roman Republic: A Reevaluation. *American Journal of Philology* 138, 143–80.

Thompson, L. A., 1962. The Relationship between Provincial Quaestors and Their Commander-in-Chief. *Historia* 11, 339–55.

Thomsen, R., 1980. *King Servius Tullius: A Historical Synthesis*, Copenhagen.

Thornton, J., 2013. Polybius in Context: The Political Dimension of the *Histories*.

In: Gibson, B., Harrison, T. (eds.), *Polybius and His World: Essays in memory of F. W. Walbank*, Oxford, 213–29.

Torelli, M., 1982. *Typology and Structure of Roman Historical Reliefs*, Ann Arbor.

Toynbee, A. J., 1965a. *Hannibal's Legacy: The Hannibalic War's Effects on Roman Life, Volume I: Rome and Her Neighbours before Hannibal's Entry*, London.

Toynbee, A. J., 1965b. *Hannibal's Legacy: The Hannibalic War's Effects on Roman Life, Volume II: Rome and Her Neighbours after Hannibal's Exit*, London.

Tränkle, H., 2009. Livy and Polybius. In: Chaplin, J. D., Kraus, C. S. (eds.), *Livy*, Oxford, 476–95.

Tucci, P. L., 2005. 'Where High Moneta Leads Her Steps Sublime': The '*Tabularium*' and the Temple of Juno Moneta. *Journal of Roman Archeology* 18, 6–33.

Van der Meer, L. B., 2007. Liber Linteus Zagrabiensis. *The Linen Book of Zagreb: A Commentary on the Longest Etruscan Text*, Dudley.

VanDerPuy, P., 2020. The Price of Expansion: Agriculture, Debt-Dependency, and Warfare during the Rise of the Republic, c. 450–287. In: Armstrong, J., Fonda, M. P. (eds.), *Romans at War: Soliders, Citizens, and Society in the Roman Republic*, London, 35–51.

Vervaet, F. J., 2014. *The High Command in the Roman Republic: The Principle of the* Summum Imperium Auspiciumque *from 509–19 BCE*, Stuttgart.

Walbank, F. W., 1957. *A Historical Commentary on Polybius: Volume I, Commentary on Books I–VI*, Oxford.

Walbank, F. W., 1971. The Fourth and Fifth Decades. In: Dorey, T.A. (ed.), *Livy*, Edinburgh, 47–72.

Walbank, F. W., 1972. *Polybius*, London.

Walbank, F. W., 2002a. Polybius as Military Expert. In: Hill, P.R. (ed.), *Polybius to Vegetius: Essays on the Roman Army and Hadrian's Wall Presented to Brian Dobson to Mark His 70th Birthday*, Privately Published, 19–30.

Walbank, F. W., 2002b. *Polybius, Rome and the Hellenistic World: Essays and Reflections*, Cambridge.

Walsh, P. G., 1970. *Livy: His Historical Aims and Methods*, Cambridge.

Walsh, P. G., 1994. *Livy Book XXXIX (Liber XXXIX) Edited with an Introduction, Translation & Commentary*, Warminster.

Waterfield, R. (trans.), 2010. *Polybius: The Histories*, Oxford.

Watson, G. R., 1956. The Pay of the Roman Army: Suetonius, Dio and the *Quartum Stipendium. Historia* 5, 332–40.

Watson, G. R., 1958. The Pay of the Roman Army. The Republic. *Historia* 7, 113–20.

Weigel, R. D., 1986. Meetings of the Roman Senate on the Capitoline. *L'Antiquité Classique* 55, 333–40.

Weinstock, S., 1971. *Divus Julius*, Oxford.

White, K. D., 1970. *Roman Farming*, London.

Wilkes, J., 2001. The Pen behind the Sword: Power, Literacy and the Roman Army. *Archaeology International* 5, 32–35.

Williamson, C., 1987. Monuments of Bronze: Roman Legal Documents on Bronze Tablets. *Classical Antiquity* 6, 160–83.

Wiseman, T. P., 1970. The Definition of '*Eques Romanus*' in the Late Republic and Early Empire. *Historia* 19, 67–83.

Woolf, G., 2009. Literacy or Literacies in Rome? In: Johnson, W. A., Parker, H. N. (eds.), *Ancient Literacies: The Culture of Reading in Greece and Rome*, Oxford, 46–68.

Ziolkowski, A., 1992. *The Temples of Mid-Republican Rome and Their Historical and Topographical Context*, Rome.

Index

Note: Roman names are listed by *nomen* unless they have a common modern variant. All dates are BC. Page numbers in *italics* refer to figures.